Emeritus Professor Arthur Roy Clapham

The Flora and Vegetation of Britain

Origins and Changes – The Facts and their Interpretation

Edited by J. L. Harley and D. H. Lewis

A Symposium to honour the 80th birthday of
Emeritus Professor Arthur Roy Clapham
CBE, FRS, FLS, MA, PhD, Hon DLitt
held at the University of Sheffield on 19 May 1984

(Reprinted from *The New Phytologist*, **98**, 1, 1984)

Published for the New Phytologist Trust

by ACADEMIC PRESS

LONDON   ORLANDO   SAN DIEGO   NEW YORK
TORONTO   MONTREAL   SYDNEY   TOKYO

ACADEMIC PRESS INC. (LONDON) LIMITED
24/28 Oval Road
London NW1
(Registered Office)

US edition published by
ACADEMIC PRESS INC.
111 Fifth Avenue
New York
New York 10003

Copyright © 1984 The New Phytologist
ALL RIGHTS RESERVED
No part of this volume may be reproduced,
by photostat, or any other means without
written permission from the trustees.

(Reprinted from *The New Phytologist*, **98**, 1, 1984)

© 1985 The New Phytologist
ISBN 0 12 325570 8

LCCN 85 70727

Printed in Great Britain at the University Press, Cambridge

CONTENTS

Introduction to symposium.
    By J. L. HARLEY 1
The relation between the British and the European floras.
    By S. M. WALTERS 3
The ecology of species, families and communities of the contemporary British flora.
    By J. P. GRIME 15
Newferry and the Boreal-Atlantic transition.
    By A. G. SMITH 35
Anthropogenic changes from neolithic through medieval times.
    By G. W. DIMBLEBY 57
Post-medieval and recent changes in British vegetation: the culmination of human influence.
    By D. A. RATCLIFFE 73
Cytogenetic variation in the British flora: origins and significance.
    By T. T. ELKINGTON 101
The flora and vegetation of Britain: ecology and conservation.
    By C. D. PIGOTT 119
INDEX 129

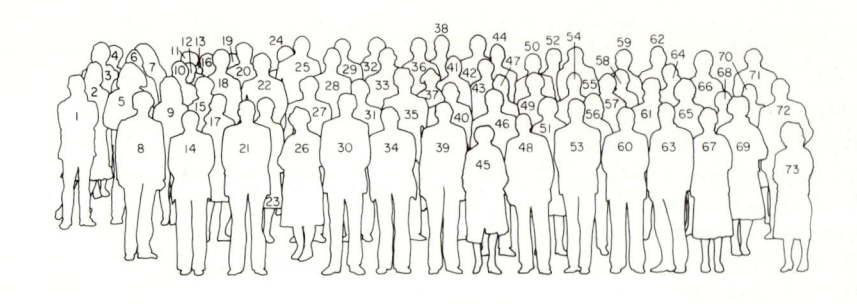

1. C N Wright
2. J P Palmer
3. B A Emmett
4. J A Lee
5. T R Pearce
6. D S Middleton
7. E Scandrett
8. D H Lewis
9. J A Crawford
10. D A Stewart
11. ?
12. A P Connolly
13. B Lee
14. T T Elkington
15. D R Peterson
16. ?
17. E M Lind
18. C P D Birch
19. V Conway
20. A Willmot
21. G W Dimbleby
22. D Briggs
23. S H Hillier
24. J S Burley
25. H W Woolhouse
26. L Walters
27. J P Grime
28. C D Pigott
29. B A Thomas
30. S M Walters
31. N J Dix
32. J G Hodgson
33. M H Martin
34. R G West
35. M M Yeoman
36. R Hunt
37. M Green
38. A F S Taylor
39. A J Willis
40. P J Syrett
41. M E Proctor
42. S C Clark
43. D A Ratcliffe
44. M C F Proctor
45. B N CLAPHAM
46. J Webster
47. J Griffiths
48. A R CLAPHAM
49. A Redda
50. J F Hope-Simpson
51. M E Webster
52. M J Russell
53. I H Rorison
54. J O Rieley
53. B D Wheeler
56. C T Williams
57. D J Read
58. S E Page
59. S C Shaw
60. A G Smith
61. P J Newbould
62. A J M Baker
63. R Watling
64. G A F Hendry
65. M Rorison
66. J C Rose
67. P P C Warburg
68. M J Loukes
69. L Harley
70. D M Holland
71. D C Smith
72. J L Harley
73. R E D Cook

Among those attending the Symposium but not present for the photograph were: Sir Harry and Lady Godwin, E Barron, T B Boam, G and B Fearn, S Lloyd, D M Lösel and J M and Mrs Thoday.

Participants of Symposium (key opposite)

THE FLORA AND VEGETATION OF BRITAIN: ORIGINS AND CHANGES – THE FACTS AND THEIR INTERPRETATION

A Symposium to honour the 80th birthday of
Emeritus Professor Arthur Roy Clapham
CBE, FRS, FLS, MA, PhD, Hon DLitt
held at the University of Sheffield on 19 May 1984

Edited for *The New Phytologist Trust*
By J. L. Harley and D. H. Lewis

INTRODUCTION TO SYMPOSIUM

By J. L. HARLEY

The Orchard, Old Marston, Oxford OX3 0PQ, UK

The following seven papers were presented at a meeting to celebrate the 80th birthday of our friend and colleague, Professor A. R. Clapham, who had from the beginning the distinction as a biologist of sharing a birthday with Linnaeus. Clapham's interest in the scientific study of plants began, indeed, in his early years and, although his reputation is worldwide as an ecologist and taxonomist, he has a deep knowledge of, and continuing interest in, the whole field of biology. After his two firsts and other student achievements at Cambridge, his first researches were in plant physiology under the great F. F. Blackman. Later, at Rothamsted, R. A. Fisher founded his interest and skill in the application of statistics to biological problems. At Oxford, as a University Demonstrator between 1930 and 1944, his association with A. G. Tansley, H. Baker, W. O. James and T. G. B. Osborn broadened his interests into morphology, taxonomy and, of course, ecology. It was during this period that his close association with Harry Godwin sharpened his interest in the history of British vegetation and the British flora which is the subject of this symposium.

In ecology and taxonomy, his contributions have given him the greatest pleasure and the highest reputation. The *Check List of British Vascular Plants* and *The Biological Flora of the British Isles* were his brain children. In cooperation with T. G. Tutin and E. F. Warburg, the various editions of *Flora of the British Isles* and the *Excursion Flora* were produced. Their publications represented a turning point in the study of plant life in Britain and had a powerful influence on the production of *Flora Europaea* as the later papers will show.

These achievements should have been enough for any man but Roy Clapham also played a leading part in Nature Conservation, in the International Biological Year and in the establishment of biology in Universities at home and abroad. In addition many present at the Symposium were taught by him at Oxford or Sheffield and all present were influenced in their work and thinking about plant sciences by him. That is not all; the University of Sheffield owes him a debt not only for building up a great Department of Botany, but also for his work on Academic Committees and as Acting Vice-Chancellor in the inter-regnum of 1965. I shall not detail the public honours that he has received, for the papers which follow by close colleagues or his former students are a greater tribute to his influence on botany, by teaching and research, by personal discussion and by editing *The New Phytologist* for 30 years.

THE RELATION BETWEEN THE BRITISH AND THE EUROPEAN FLORAS

By S. M. WALTERS

University Botanic Garden, Cambridge CB2 1JF, UK

SUMMARY

The paper presents a brief and personal history of the main developments in British vascular plant taxonomy and distribution studies in a European setting since the Second World War. It assesses the importance of the 'European view', and considers recent and current work under two main heads: the factual information itself, and the presentation of the facts, especially in the form of dot distribution maps.

Key words: British vascular plants, European vascular plants, distribution studies, dot distribution maps.

INTRODUCTION

I was delighted to receive the invitation to contribute to this Symposium in honour of Professor Clapham on his 80th birthday, although I was naturally at the same time sorry that Professor Tutin, on every possible criterion the person who should be giving this particular introductory address, had had to decline on health grounds. I suppose I could be said to belong to the 'first C.T. & W. generation' of botanists in Britain: my school Flora was 'Bentham & Hooker' with an occasional reference to other works, including the incomparable Hayward's Botanists' Pocket Book (Druce, 1930), and at University Humphrey Gilbert-Carter's influence made sure that it was Hooker's Students' Flora (1870) that I used. As Humphrey stated very clearly in the Preface to his charming little booklet *Catkin-bearing Plants* (1930), this was 'the only Flora of this country fit to be put into the hands of a learner'. I was never very clear where he rated Babington's *Manual of British Botany* in its 10th (1922) edition by A. J. Wilmott, but increasingly that was the Flora I used in my post-war Cambridge years, when we were all waiting for Clapham, Tutin & Warburg (1952)!

I mention something of my own botanical education here because it enables me to stress how much I owe to those influences which from the first prevented me from being a narrowly insular taxonomist and in particular took me, long before Mr Heath and Mrs Thatcher, 'into Europe'. The first influence is particularly relevant to recall on this return visit to Sheffield, the 'big city' of my childhood. In the last year of my schooldays at Penistone Grammar School, the West Riding County Council, as it then was, offered for the first (and I believe, because of the outbreak of war, the *only*) time, a number of travelling scholarships. I was fortunate enough to win one, and went to spend an idyllic and greatly rewarding summer in the Alpine Garden of the University of Geneva at Bourg St Pierre. It was my first trip abroad, my first sight of real mountains, my first real linguistic exercise.

Note: throughout this paper the term 'British Isles' is used in its pure geographical sense to include Ireland.

It made an indelible impression, which my Cambridge career, with its wide international scientific view, simply used and developed.

In a very special sense, 'C.T. & W.' is a Humphrey Gilbert-Carter product. Tom Tutin tells the story in his contribution to the Gilbert-Carter Memorial Volume (Gilmour & Walters, 1975) and Roy Clapham in the Tutin 70th Birthday Festschrift (Clapham, 1978a); do read them if you don't already know the story of the 'Tansley tea'. I count myself especially privileged to have been a 'junior partner' in the famous collaborative British Flora team: and in particular I acknowledge my gratitude to E. F. Warburg, whose early death we all greatly lamented, for it was 'Heff' who really sparked my enthusiasm for 'critical groups' in the British flora in general, and for *Alchemilla* in particular.

THE EUROPEAN VIEW

Why was this European view so necessary in British taxonomic studies? One reason is abundantly clear in the history of the Botanical Society of the British Isles (BSBI) Distribution Maps Scheme, in the preparation of which Professor Clapham played the leading part. It was Clapham who saw in 1950 how much we could usefully learn in British botany from the work and the experience of botanists in other European countries and, in the event, our Maps Scheme borrowed ideas from both Scandinavia, where Hultén's *Atlas* (1950) had pioneered the field, and from Holland, where a printed field record-card was already in use. The introduction to the *Atlas of the British Flora* (Perring & Walters, 1962) explains these important Continental European influences. It was not only in taxonomy and the traditionally-linked study of species distributions that Clapham saw the value of a European view. I recall, for example, his characteristically stimulating contribution to the 1952 Conference of the Botanical Society on *The Changing Flora of Britain* (Clapham, 1953), in which he considered carefully the evidence for the native status of five widespread British plants, explaining that his curiosity in this matter had been aroused 'during a visit to Sweden in the Summer of 1950', when he had seen all five species growing 'as constituents of little-disturbed natural vegetation'.

There is, of course, nothing new in British botanists studying their own flora in a European context. As William Stearn makes clear in a valuable essay on the British contribution to the study of the European flora (Stearn, 1975), the 'European view' we have now is more akin to that of the pioneers in the study of British plants, Turner in the sixteenth century and Ray in the seventeenth, than it is to the botanists of the Victorian period, when so much interest was naturally directed outside Europe to the floras of parts of the British Empire in particular. So our current revived interest can be seen as part of a general cultural and historical swing – a 'post-imperialist' attitude, in fact.

For taxonomists, the culmination of the process of 'going into Europe' has been the completion of the five-volume standard work *Flora Europaea* (Tutin *et al.*, 1964–1980): in this fine collaborative enterprise, chaired by Tutin, Clapham was not only an author, but was also one of the team of Advisory Editors drawn from different European countries. I must resist the temptation to turn this talk into an anecdotal account of the *Flora Europaea* project, though the temptation to do so is very great indeed: for what better example can we have of the practical relation between the British and European floras than a completed standard work in which 'Br' takes its humble place among 'Ga', 'Hs', 'No', etc. (standard country abbreviations indicating broad European distribution)? It is very salutary indeed to be reminded, for example, that one of our rarest and most treasured British natives, the orchid *Cephalanthera rubra*, occurs throughout 'most of Europe, northwards to S. England and S. Finland' – or that, conversely, the Bluebell (*Hyacinthoides nonscripta*) is native outside the British Isles only in France, the Low Countries and the Iberian peninsula. A European view of the British flora inevitably alters our priorities, not least in the very urgent and practical questions of nature conservation. On an international scale, the conservation of a range of bluebell woods could well be a particular task for the British Nature Conservancy, if we take a long view of our responsibilities.

There is one other field in which the present generation of keen students of the British and European floras is enormously better equipped than was my generation. I refer to Quaternary studies, and especially to Sir Harry Godwin's standard work on the *History of the British Flora* (Edn 2, 1975). The whole development under Godwin of palynology and other sub-fossil studies of past species distributions has been from the first a European, not a parochially British, study, and the most recent publication in this field (Huntley & Birks, 1984), which traces through palynology the postglacial history of individual genera and species across the face of Europe, puts into the hands of the field botanist a set of data of quite extraordinary interest. We can now, for example, see in a series of maps the stages over some 12000 years by which the present distribution of the Crowberry (*Empetrum*) has been achieved, or the detailed change in the pattern of native Beech (*Fagus*) as it colonizes northwards and westwards, reaching S. England some 3000 years ago. Like many other field botanists, I am sampling with pleasure these extraordinary maps, and enjoying the explanatory text accompanying each mapped genus. It is abundantly clear, as Godwin claims in the Preface to the *History of the British Flora*, that 'an ounce of historical geological fact is worth a ton of historical speculation...based on nothing more than comparison of present-day distributions'. It is nevertheless true that biologists will go on speculating about the explanation of present-day patterns of plant and animal distribution, although one hopes that all such speculation will always include relevant evidence from Quaternary studies. There is a clear analogy here with the speculation of biologists interested in evolutionary processes on the significance of present-day taxonomic patterns: it is surely incumbent upon them to take into account any relevant fossil evidence. A particularly interesting and careful re-assessment of the significance of distribution patterns of plant species in their European setting has just been published by Professor David Webb for the Irish flora (Webb, 1983): we await a similar exercise to correct and update the literature on the British flora, and I fear that some in my audience will be disappointed that this paper can only sketch the outline of such a study.

Geographical Relationship of the British Flora

The starting-point for any such study must obviously be the work of Matthews, whose earliest papers on the relationship of the British flora to that of Europe and the world in general appeared in 1923. His most-quoted work was the Presidential Address he gave to the British Ecological Society in 1935 entitled *Geographical relationships of the British flora* (Matthews, 1937), and his ideas are conveniently available in an undeservedly neglected student text-book published in 1955. It is this book, now 30 years old, which deserves a critical revision. I appreciate the opportunity this lecture affords me to pay tribute to Matthews' work: so far as my own interest in the origin of the British flora is concerned, the 1937 paper was an essential reference point, and it was a particular pleasure to me, and to my colleague Frank Perring, to be able to include in the *Atlas of the British Flora* (1962) a Foreword by Professor Matthews himself.

Matthews' approach was directly linked to that of the great nineteenth century pioneers in the study of plant and animal distribution, Forbes, Watson, Hooker, Wallace and of course Darwin himself. It is worth re-reading what Matthews wrote in the introduction to his 1924 paper, the second in a series which considers the significance of particular elements in the British flora. He says:

"A close relationship invariably exists between the flora of a continental island and that of the nearest mainland, a good example being provided by the British flora itself. For a general discussion of such island floras, their origin and distribution, one need only turn to Darwin's 'Origin of Species' or Wallace's 'Island Life'. In these classics the theory of migration is advanced as a sufficient and satisfactory explanation of the known facts of distribution. Darwin's views are strikingly summarized in the following sentence: 'If the difficulties be not insuperable in admitting that in the long course of time all the individuals of the same species, and likewise the several species belonging to the same genus, have proceeded from some one source, then all the grand leading facts of geographical distribution are explicable on the theory of migration, together with subsequent modification and the multiplication of new forms . . . .' Whatever the future verdict regarding certain outstanding difficulties, the theory of migration must always stand as accounting for all the 'leading facts' of plant-geography. It can never be seriously challenged, least of all when a comparatively small part of the earth's surface such as Britain, with its known geological history, is the area under consideration. For the flora of Britain is almost wholly an outpost of the larger European flora, and, fragmentary though the fossil record be, there is reason to believe that our native flora has attained its present range in the country as a result of past migrations brought about largely by climatic change. With the historical factor becoming more clearly defined, these distribution studies were commenced in the hope that some light might be shed on the several plant invasions from the Continent which have shared in the building up of the flora of Britain, and to ascertain, also, how far the 'Age and Area' theory of distribution propounded by Dr Willis might be applicable to our rather complex flora." (see Willis, 1922.)

I must resist the temptation to talk about J. C. Willis and the 'Age and Area' hypothesis: it is a fascinating subject, but takes us too far from our main theme. What I *do* want to consider may be brought together under two heads, namely: the factual information itself; and its presentation in cartographic or other form.

Factual Information on the British Flora in a European Setting

I have already indicated the enormous advances in our knowledge concerning the former distribution of British species. It is, however, salutary to bear in mind that much of this information is accurate to the *generic* level only, and some of it, like the palynological record of sedges or grasses, only at the level of the family. Where over much of present-day Europe we are dealing with a single species of the particular genus, it may be entirely justified to assume that sub-fossil pollen and modern populations refer to that same species – this is the case with *Fagus sylvatica*, the Beech, even though *Fagus orientalis* is also a European native.

However, most cases are not so simple, and we must accept that there is a graded series here involving subjective decisions as to what it is reasonable and unreasonable to assume. Partly this is a problem of taxonomic discrimination using pollen morphology, but partly it remains a question of judicious assessment. I mention this because, although I agree entirely with Godwin's insistence that 'an ounce of fact is worth a ton of speculation', I think it is too simplistic to see the factual information as supremely hard and irrefutable. What we *can* and must say, however, is what the 'facts' are as we know them. Thus, to take a familiar example, the use of sub-fossil data on the distribution of *Helianthemum* is severely limited by the uncertainty attaching to the discrimination of the species, and Watts (1983) has recently written, with regard to the remarkable Burren flora: 'There is little in the pollen record to spell out the history of the species which give the Burren its floristic interest today...Pollen of *Helianthemum* spp. occurs sporadically throughout the post-glacial; one tends to assume it is of *H. canum*, but it is possible that *H. nummularium*, now confined to a single station in Co. Donegal, might have been present too'. [Godwin (1975) gives a full discussion of the problem: pp. 139–141.]

Nor is it entirely safe to assume that modern taxonomic judgment is free from subjective assessment, or that what C.T. & W. (Edn 2) or *Flora Europaea* state is a 'hard fact'. Quite apart from those notorious cases where generic limits are different in the two standard works (there are some glaring discrepancies in Gramineae and Cyperaceae, for example), the continuing process of taxonomic study produces important additions to our knowledge of the relationships between British and European plants, even in genera not usually considered particularly difficult or (in the taxonomic sense) critical. A good example is *Myosotis brevifolia*, an apparently endemic species described by C. E. Salmon in 1926, and so treated in C.T. & W. Edn 2, which in the account by Grau & Merxmuller in *Flora Europaea* (**3**: 115) is 'sunk' in the W. European *M. stolonifera*.

Within the critical apomictic genera, of course, new discoveries are more likely. Perhaps I can illustrate this by reference to the critical genus I know best – *Alchemilla* – which affords an excellent example of how a knowledge of the European flora was, correctly, predictive of what we would eventually find in Britain. During the early 1940's, when Warburg & Wilmott were collaborating (insofar as the Second World War permitted) to elucidate both *Alchemilla* and *Sorbus* taxonomy in Britain – which is where I came in! – their *Alchemilla* discoveries were just beginning in Teesdale, and they had begun to suspect that there was more than one 'continental' *Alchemilla* species in the remarkable Upper Teesdale 'assemblage'. By 1948 we had 'discovered' there, in addition to *A. monticola* (known, as *A. pastoralis*, in the 1920's), two other widespread European species: *A. acutiloba* and *A. subcrenata*.

This history is summarized, and its phytogeographical significance considered, by Clapham himself in his editorial introduction to the book entitled *Upper Teesdale: the area and its natural history* (Clapham, 1978b, pp. 20–21). What is not mentioned there is that Margaret Bradshaw (as she then was – now Margaret Proctor) and I both continued to expect to find, somewhere in the Teesdale area, the fourth *Alchemilla* species with a widespread European distribution similar to these three – namely *A. gracilis* – but it eluded us...until 1976, when it was found in Northumberland by Professor G. A. Swan, and identified and verified by me in 1978. (I have to confess that this remarkable addition to the British flora has still not been 'written up' in detail.)

Another important aspect of factual information is, of course, floristic *change*. In the negative sense – the decline and even the local extinction of species – of course this is part of the urgent and serious concern of the nature conservation movement, and I refrain from saying more about this beyond the obvious statement, namely that our published information is enormously greater now (The 'Red Data Book', Perring & Farrell, 1983, is an obvious case). But in the positive sense – the recording of spread of native or, more likely, introduced taxa – the vigilance and industry of the members of the Botanical Society of the British Isles in particular continues to be, as it has been for many years, of very special importance. Again, this is a study in which a European, even a world, dimension is essential. Perhaps I can illustrate by a piquant example.

This concerns the North American duckweed, *Lemna minuscula*. In September 1977 we celebrated in Cambridge the completion of the text of the fifth and final volume of *Flora Europaea*. It was a happy occasion, and we were particularly pleased that a good number of our Continental colleagues who were Regional Advisers, authors, or both, were able to be present to join in the celebrations. One of those present was Professor Elias Landolt from Zurich, the Regional Adviser for Switzerland (He). Strolling back one evening from the Botanic Garden to King's College, he collected a sample of *Lemna* from the ditch by the River Cam on Coe Fen, took it home with him, grew it and confirmed its identification as the American species *Lemna minuscula*. He published this record, the first for the British Isles, in 1979. Once alerted, British field botany swung into action, so that by the end of 1981 the alien duckweed had been recorded from no fewer than eleven vice-counties. It was obviously an overlooked plant. Note that with 'new' alien species, the literature is soon out of date. Not even the recent (1981) third edition of the C.T. & W. *Excursion Flora* contains *Lemna minuscula*, and though Volume 5 of *Flora Europaea* (1980) includes it, as naturalized in France, it appears under the name of the closely-allied N. American species *L. valdiviana*. Landolt's detailed work on the family was published in 1980, and Alan Leslie and I published a short paper on the species in Britain last year in *Watsonia* (Leslie & Walters, 1983). An example of European taxonomic co-operation where the British are for once the receivers of specialist advice about their own flora! I hasten to say that not all such advice from the European continent is quite so readily taken up in these islands. Webb, in the paper I cited earlier, is firmly dismissive of the claims to the status as endemic Irish species of no fewer than three taxa described by Continental botanists in recent years, namely: *Antennaria hibernica*, *Rumex hibernicus* and *Salix hibernica*. The *Antennaria* comes off worst: it is summarily disposed of as 'a figment of Braun-Blanquet's phytosociological enthusiasm'!

The Presentation of Facts and Figures

There are two main ways in which the facts of plant and animal distributions may be assembled for use in argument: as statistics or as maps. Willis made great play with the statistical presentation of phytogeographical data, and much of interest arises from his work and similar more recent studies. I have, however, set my face against any detailed assessment of this kind of study for the purposes of this lecture, and will content myself with the observation that there is an element of subjective uncertainty in the first stages of assembling any statistics relevant to phytogeography. This emerges clearly from Webb's critique (as it applies to the Irish flora) of some of Matthews' decisions about the position of individual species in his so-called 'geographical elements'. Webb demurs, rightly, at Matthews' inclusion of *Saxifraga hirsuta* in the 'alpine element' of the British flora, for example: and I could add my own cases where it seems Matthews simply got the facts wrong – *Myosotis alpestris*, for example, is *not* an arctic–alpine species. Assessing the endemic element in our flora is particularly prone to subjective assessment: I struggled with this problem in a paper for Professor Tutin's 'Festschrift' (Walters, 1978), and have to confess that I found the result only tolerably satisfactory. At least it concentrated my mind on the problems of taxonomic assessment and use of phytogeographical data!

I should like to spend a little more time on the other main method of presentation, namely mapping. This subject is, of course, particularly appropriate to the occasion for, as I have already reminded you, it was Roy Clapham who, as Secretary of a special committee appointed by the Botanical Society of the British Isles in 1950, prepared the detailed case which set up the Distribution Maps Scheme in 1953. Indeed, I can say without exaggeration that my whole career was shaped by the very rewarding association I, as Director of the Scheme, had with Roy as Secretary of the Maps Committee. We were early convinced of the value of an objective presence-or-absence recording system based upon the Ordnance Survey National Grid: we saw that this could be combined with an automated retrieval system, and in the event, Frank Perring and I were able to pioneer the use of mechanically sortable punched cards to make dot-maps automatically. It is, incidentally, incredible how rapidly technological changes have come and gone – our famous punched-card sorter and tabulator used in the 1950's to produce the Atlas maps are now on show in a *Computer Museum* in Massachusetts!

In 1956 I presented a paper to the B.S.B.I. Conference of that year entitled *Distribution Maps of Plants – a Historical Survey* (Walters, 1957), and finished the lecture with these words: 'Sooner or later, a sufficient number of people will be interested in a cooperative project to map the European flora. Perhaps here we can repay our debt to European phytogeography by suggesting a practicable, standardized method for doing it. I have ventured to outline such a method as part of our Maps Exhibit at this Conference.'

Nearly 30 years later, we have a remarkably effective *Atlas Florae Europaeae* project with a Secretariat in Helsinki, Finland, under Professor Jaakko Jalas; and although progress in publishing the Atlas remains slow (it was recently said at an Atlas Committee Meeting that to finish mapping the European flora at the present rate of progress would take another 80 years!), nevertheless six volumes are already published and the seventh and eighth are in active preparation.

The Atlas maps are based upon the 50 km grid square of the Universal Transverse Mercator (UTM) projection maps which cover the whole of Europe as defined for *Flora Europaea*...a system which my colleague Frank Perring pioneered. They are beautifully prepared and printed, and the whole series deserves to be much more widely known. (I am pleased to say that negotiations are now taking place between the Finnish 'Vanamo' Society and Cambridge University Press with the aim of improving the sale of the Atlas in W. Europe in particular.) To illustrate their quality and, I hope, their value to anyone interested in British and European phytogeography, I have selected four maps of species of the genus *Sagina* published in the sixth and most recent volume (Jalas & Suominen, 1983). *Sagina* is especially appropriate because Clapham was joint author (with Jardine) of the account in the first volume of *Flora Europaea* (1964).

Sagina subulata

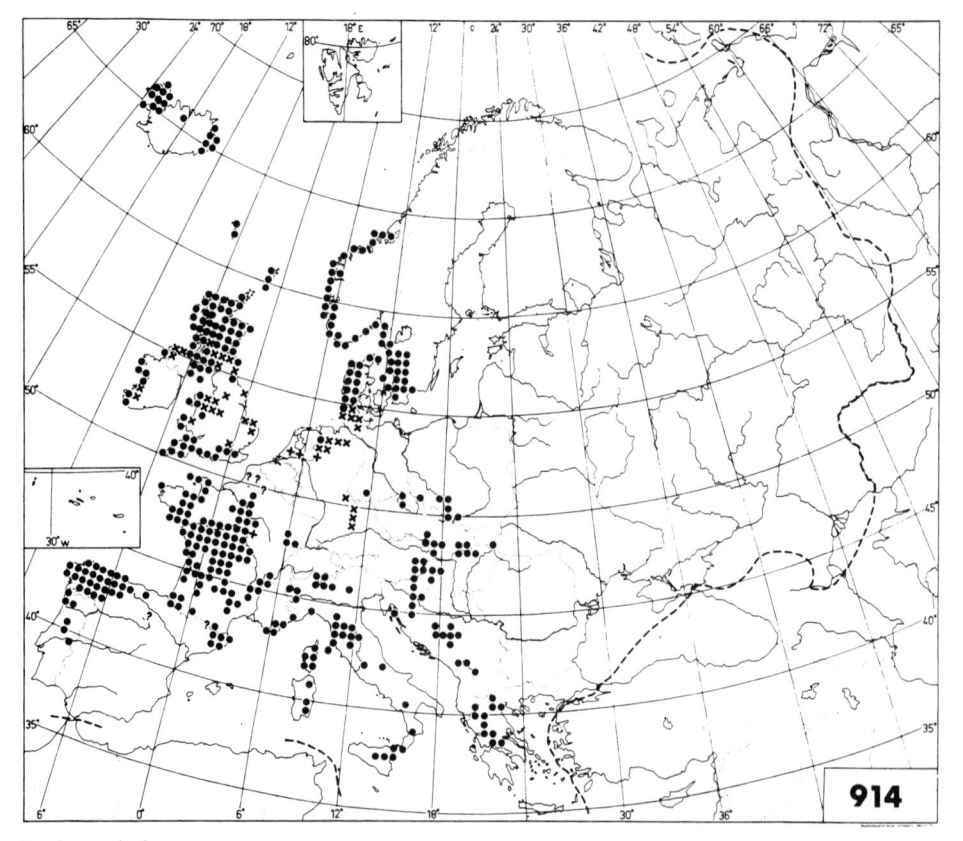

Sagina subulata (Swartz) C. Presl – Map 914.

Spergula subulata Swartz. Incl. Sagina revelieri Jordan & Fourr.; S. saginoides (L.) Karsten var. revelieri (Jordan & Fourr.) Fiori

Notes. Be omitted (given in Fl. Eur.). Presumably extinct in Ho (given as present in Fl. Eur.): Atlas Nederl. Fl. 1980: 176. The records listed by T. Săvulescu (ed.), Fl. Reipubl. Pop. Romanicae 2: 75 (Bucureşti 1953), possibly belong to *Sagina nodosa*.

Sagina sabuletorum (Gay) Lange – Map 915.

Sagina loscosii Boiss.; Spergula sabuletorum Gay

Notes. Also recorded for Hu (not given in Fl. Eur.), as *Sagina saginoides* subsp. *macrocarpa* (Reichenb.) Soó var. *karolyiana* Soó, Bot. Közlem. 45: 264 (1954), the taxon being later synonymized with *S. sabuletorum* by A. Pénzes, Savaria 2: 52 (1964). Perhaps an alien in Hu, see Soó Synopsis 1970: 358.

Total range. Outside Europe, present in Morocco (given as endemic to Europe in Fl. Eur.): R. Maire, Fl. Afr. Nord 9: 239 (1963).

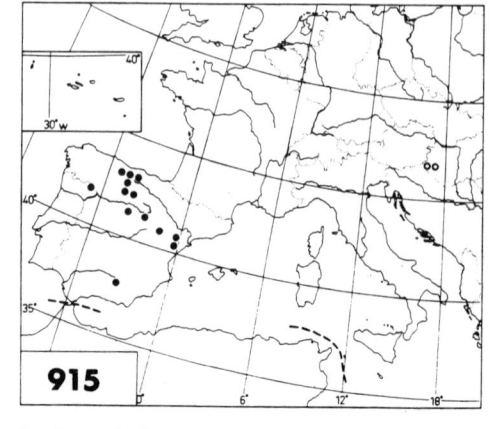

Sagina sabuletorum

Fig. 1. The distribution of five *Sagina* spp. in Europe. [Maps 914–916 and 920 from *Atlas Florae Europaeae* (Jalas & Suominen, 1983), on which each dot represents presence in a 50 km grid square.]

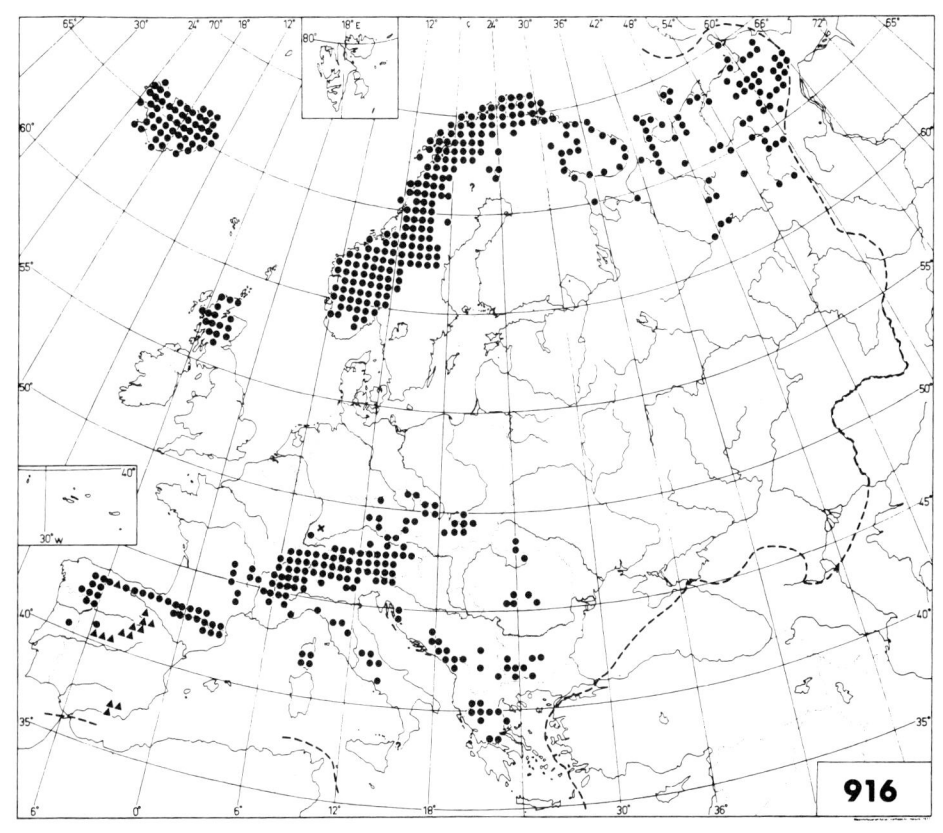

● = Sagina saginoides   ▲ = S. nevadensis

Sagina nevadensis Boiss. & Reuter – Map 916.

Sagina saginoides (L.) Karsten var. glandulosa (Lange) Rivas Martínez; S. saginoides var nevadensis (Boiss. & Reuter) Briq.; S. saginoides subsp. nevadensis (Boiss. & Reuter) Greuter & Burdet

Nomenclature. W. Greuter & T. Raus (eds.), Willdenowia 12 (Cahiers Optima Leaflets 127): 189 (1982).

Notes. Lu added (not given in Fl. Eur.): S. Rivas Martínez & C. Saenz de Rivas, An. Real Acad. Farmacia 45: 595–596 (1979); S. Rivas Martínez, An. Jardin Bot. Madrid 36: 306 (1979).

Total range. Outside Europe, recorded from Grand Atlas in Morocco (given as endemic to Europe in Fl. Eur.): R. Maire, Fl. Afr. Nord 9: 237 (1963).

Sagina saginoides (L.) Karsten – Map 916.

Alsine linnaei (C. Presl) Jessen; A. saginoides (L.) Crantz; Arenaria frigida Rupr.; Sagina linnaei C. Presl; S. olympica Stoj. & Jordanov; S. rosonii Merino; S. saxatilis (Wimmer & Grab.) Wimmer; S. spergella Fenzl; Spergella saginoides (L.) Reichenb.; S. saxatilis (Wimmer & Grab.) Schur; Spergula saginoides L.; S. saxatilis Wimmer & Grab. Incl. Sagina macrocarpa (Reichenb.) J. Maly; S. saginoides subsp. macrocarpa (Reichenb.) Soó; S. saginoides var. macrocarpa (Reichenb.) Moss; Spergella macrocarpa Reichenb. Excl. Sagina x normaniana Lagerh.; S. procumbens x saginoides; S. saginoides subsp. scotica (Druce) Clapham

Taxonomy. G. E. Crow, Rhodora 80: 34–42 (1978).

Notes. Present in Sa according to Pignatti Fl. 1982: 224, but no exact data available (not given in Fl. Eur.).

Total range. Hultén Alaska 1968: 426; Hultén CP 1971: map 44.

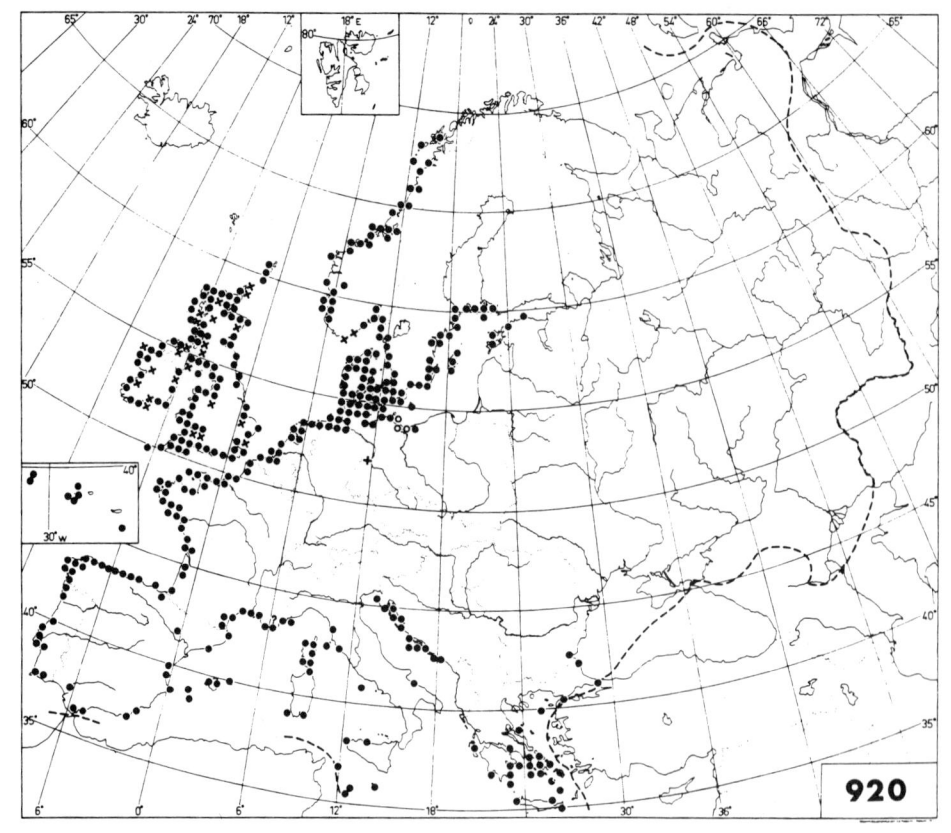

Sagina maritima

Sagina maritima G. Don – Map 920.

Alsine donii C. F. W. Meyer; A. maritima (G. Don) Jessen; Sagina apetala Ard. var. maritima (G. Don) Wahlenb.; S. filiformis Pourr.; S. rodriguesii Willk.; S. stricta Fries; S. urceolata Viv.

Notes. Az, Ge, Si and Tu added (not given in Fl. Eur.): Fl. Turkey 1967: 91; Pignatti Fl. 1982: 225. *Total range.* MJW 1965: map 146d.

In the 20 years since the F.E. account was published, 'new' information of all kinds has been accumulated, and some of this is taken into account in publishing the Atlas maps. Thus, for *Sagina maritima*, records for no fewer than four European countries not given in F.E. have been incorporated in this map (Az, Ge, Si & Tu).

The *Flora Europaea* Committee has recently appointed Dr John Akeroyd as Research Officer to assemble and assess all such *addenda* and *corrigenda* for the first volume of the Flora with a view to eventual publication of a second edition. Coordination with the *Atlas* project, and also with the European Taxonomic, Floristic and Biosystematic Documentation System based at Reading University under Professor Heywood, is a very necessary part of this work. It is good to know that there is both the money (from the F.E. Trust Fund of the Linnean Society) and the expert interest to tackle these 'second generation' problems on a European scale, with modern methods of data processing where they are appropriate, but with the careful attention to accurate detail which has always been characteristic of the best work in European floristic and taxonomic studies.

No-one has exemplified more effectively how much science needs these qualities than Professor Clapham himself; it is his contribution to our integrated European scientific outlook that we honour today.

REFERENCES

CLAPHAM, A. R. (1953). Human factors contributing to a change in our flora: the former ecological status of certain hedgerow species. In: *The Changing Flora of Britain* (ed. by J. E. Lousley), pp. 26–39. Botanical Society of the British Isles Conference Report, Oxford.
CLAPHAM, A. R. (1978a). Thomas Gaskell Tutin. In: *Essays in Plant Taxonomy* (ed. by H. E. Street), pp. ix–xiv. Academic Press, London.
CLAPHAM, A. R. (ed.) (1978b). *Upper Teesdale: the Area and its Natural History*. Collins, London.
CLAPHAM, A. R., TUTIN, T. G. & WARBURG, E. F. (1952). *Flora of the British Isles*. University Press, Cambridge.
DRUCE, G. C. (ed.) (1930). *Hayward's Botanists' Pocket-Book* (19th edn). Bell, London.
GILBERT-CARTER, H. (1930). *Our Catkin-Bearing Plants*. University Press, Oxford.
GILMOUR, J. S. L. & WALTERS, S. M. (eds). (1975). *Humphrey Gilbert-Carter: a Memorial Volume*. University Botanic Garden, Cambridge.
GODWIN, H. (1956: Edn 2, 1975). *History of the British Flora*. University Press, Cambridge.
HOOKER, J. D. (1870). *The Students' Flora of the British Islands*. Macmillan, London.
HULTÉN, E. (1950). *Atlas över växternas utbredning i Norden*. Generalstabens Litografiska Anstalts Förlag, Stockholm.
HUNTLEY, B. & BIRKS, H. J. B. (1984). *An Atlas of Past and Present Pollen Maps for Europe: 0–13,000 Years Ago*. University Press, Cambridge.
JALAS, J. & SUOMINEN, J. (eds) (1983). *Atlas Florae Europaeae: Distribution of Vascular Plants in Europe, vol. 6 Caryophyllaceae (Alsinoideae & Paronychioideae)*. Societas Biologica Fennica Vanamo, Helsinki.
LESLIE, A. C. & WALTERS, S. M. (1983). The occurrence of *Lemna minuscula* Herter in the British Isles. *Watsonia*, **14**, 243–248.
MATTHEWS, J. R. (1924). The distribution of certain portions of the British flora. II. Plants restricted to Scotland, England and Wales. *Annals of Botany*, **38**, 707–721.
MATTHEWS, J. R. (1937). Geographical relationships of the British flora. *Journal of Ecology*, **25**, 1–90.
MATTHEWS, J. R. (1955). *Origin and Distribution of the British Flora*. Hutchinson, London.
PERRING, F. H. & FARRELL, L. (1977: ed. 2, 1983). *British Red Data Book 1. Vascular Plants*. Royal Society for Nature Conservation, Nettleham, Lincoln.
PERRING, F. H. & WALTERS, S. M. (1962). *Atlas of the British Flora*. Nelson, London.
STEARN, W. T. (1975). History of the British contribution to the study of the European flora. In: *European Floristic and Taxonomic Studies* (ed. by S. M. Walters & C. J. King), pp. 1–17. Botanical Society of the British Isles Conference Report, Cambridge.
TUTIN, T. G. et al. (eds) (1964–1980). *Flora Europaea*. 5 vols. University Press, Cambridge.
WALTERS, S. M. (1957). Distribution maps of plants – a historical survey. In: *Progress in the Study of the British Flora* (ed. by J. E. Lousley), pp. 89–96. Botanical Society of the British Isles Conference Report, London.
WALTERS, S. M. (1978). British endemics. In: *Essays in Plant Taxonomy* (ed. by H. E. Street), pp. 263–274. Academic Press, London.
WATTS, W. A. (1983). The Vegetation since the last glaciation. In: *Flora of Connemara and the Burren* (ed. by D. A. Webb & M. J. P. Scannell), pp. **4**, xxxvi–xl. University Press, Cambridge.
WEBB, D. A. (1983). The flora of Ireland in its European context. *Journal of Life Sciences of the Royal Dublin Society*, **4**, 143–160.
WILLIS, J. C. (1922). *Age and Area*. University Press, Cambridge.
WILMOTT, A. J. (ed.) (1922). *Babington's Manual of British Botany* (10th Edn) Gurney & Jackson, London.

THE ECOLOGY OF SPECIES, FAMILIES AND COMMUNITIES OF THE CONTEMPORARY BRITISH FLORA

By J. P. GRIME

Unit of Comparative Plant Ecology (NERC), Department of Botany, The University, Western Bank, Sheffield S10 2TN, UK

Summary

It is argued that detailed studies of the functioning of populations cannot by themselves lead to an understanding of British vegetation. Several complementary lines of investigation are required to explore the role of past and present interactions between habitat templet and plant characteristics. Strategy concepts provide a framework in which to assemble information from such diverse fields of ecological research, and recent results suggest that the present ecologies of populations, species, families and communities remain strongly conditioned by patterns of ecological specialization which predate the development of the post glacial flora.

Key words: British flora, strategy, angiosperm families, communities, diversity.

Introduction

In an era of increasing specialization by plant ecologists the contributions and philosophy of A. R. Clapham have been sure pointers to wider horizons and responsibilities. In particular the breadth of his interests and activities have reminded us of the essential links between taxonomy, plant ecology and the scientific management and conservation of vegetation. This symposium takes place at a time when the need for Clapham's broad and balanced approach to plant ecology has never been greater, both as an antidote to the 'narrow excesses' of Academia and as a spur to the use of ecological expertise in the world at large.

In his Presidential Address to the Ecological Society (Clapham, 1956) Clapham surmised that 'our primary concern as plant ecologists is to know why a plant of this species, and not of that, is growing in a given spot; that whatever views we might entertain about the community as organism, or quasi-organism, or as a mere assemblage of individuals, we should show ourselves to be interested primarily in autecological problems, that is in enlarging our knowledge and understanding of the biology of individual species of plants'. The paper goes on to advocate the use of controlled environment facilities to elucidate key factors in the ecology of species and (a Clapham hallmark) explains how the information might be brought together and utilized.

In this paper comment will be made on the extent to which these proposals and objectives remain valid in 1984, and an assessment will be attempted of the progress which has been achieved in understanding British vegetation as a result of reductionist approaches, of the type recommended in Clapham's paper.

Species

Species and populations

Since 1956 there has been an explosion of interest in the detailed functioning of plant populations and many insights have been gained by importing into plant ecology techniques of demography, life-history analysis and computer modelling originating from animal ecology and other branches of science. So far attention has been largely concentrated upon those populations which are particularly amenable to demographic study, although progress has been made in adapting the approaches to populations to take account of the common phenomenon (among plants) of clonal expansion. Perhaps most significant of all in these recent developments has been the establishment within the science of plant ecology of a large specialism with its own *lingua franca*, International Society and foci for publications. Our conception of the species and of plant ecology itself has been enriched by applying the theories and techniques of population biology, and it is only after careful consideration that one ventures to comment upon the potential dangers which may arise from exclusive devotion to this branch of plant ecology.

In a recent paper by a leading plant-population biologist the following passages appear:

> the search for generalities in ecology has been disappointing – more so in plant than in animal ecology. The few generalities that have emerged come from studies of stands of single species.
>
> ... It is from the work of many individuals working scattered over a variety of parts of the world, but concentrating their attention over long

Table 1. *Examples of laboratory characteristics associated with particular demographic phenomena*

Species	Laboratory characteristics	Associated demographic phenomena
Hypericum perforatum *Juncus effusus* *Milium effusum* *Origanum vulgare*	Dispersule small, compact. Germination inhibited by darkness and requiring high and/or fluctuating temperatures	Accumulation of persistent bank of buried seeds
Arrhenatherum elatius *Avenula pratensis* *Bromus erectus* *Hordeum murinum*	Dispersule long and with appendages. Germination occurs rapidly in light or dark over wide range of temperatures	Transient seed bank producing single cohort of seedlings in the autumn
Acer pseudoplatanus *Anthriscus sylvestris* *Heracleum sphondylium* *Impatiens glandulifera*	Dispersules relatively large. Germination depends on exposure to chilling	Transient seed bank producing single cohort of seedlings in the spring
Avenula pratensis *Bromus erectus* *Fritillaria meleagris* *Hyacinthoides non-scripta*	2C nuclear DNA content > 2.0 pg	Rapid leaf expansion in the early spring
Carex flacca *Carex pulicaris* *Chamerion angustifolium* *Epilobium hirsutum*	2C nuclear DNA content < 2.0 pg	Rapid leaf expansion delayed until late spring and summer
Danthonia decumbens *Helianthemum nummularium* *Nardus stricta* *Vaccinium vitis-idaea*	Relative growth rates of seedlings in productive conditions < 1.0 g g^{-1} week^{-1}	Life-span of leaves long (> 1 year)

periods on the behaviour of individual plants, that the development of
ecology as a generalizing and predictive science may be possible.

The detailed analysis of proximal ecological events is the only means
by which we can reasonably hope to inform our guesses about the ultimate
causes of the ways in which organisms behave.

Harper (1982)

In these phrases and in its dismissive treatment of alternative lines of ecological enquiry and their terminologies, Harper's paper departs from the tolerant tradition which has brought many and varied minds to bear on the British flora. Particularly harmful, in my view, is Harper's insistence upon the detailed study of proximal events in the field in contemporary populations as the only reliable way to gain a general understanding of vegetation processes. In golfing terms, this is equivalent to 'putting from the tee' and would entail (1) disregard of existing and potential research gains from studies of units which are larger and of greater antiquity and informational value than contemporary plant populations, (2) neglect of many opportunities for fruitful synthesis of evidence from complementary types of research (historical, demographic, physiological, biochemical) and (3) delay in the development of sorely needed general models and guidelines for the management and conservation of vegetation.

Table 1 provides an illustration of the scope which exists for useful syntheses of data from rather different types of ecological studies. In each of the six examples, demographic phenomena of common occurrence in British plants are related to species characteristics observed by simple laboratory measurements or experiments. In view of the time and labour required to obtain reliable field data, evidence of this kind prompts the view that laboratory screening of selected plant attributes may allow considerable economy of effort in research designed to recognize the functional characteristics of large numbers of species and populations.

Species and strategies

One of the casualties of the recent narrowing in the focus of much ecological work has been the plant *species* as an object of study. It is interesting to note from the quotation cited in the Introduction to this paper that Clapham, despite his intimate acquaintance with infraspecific variation (Clapham, Tutin & Warburg, 1952), does not hesitate to recommend the ecological study of species. We may be fairly certain that this was founded on his assumption that the species would remain the only readily recognizable vegetation unit and would continue to provide a widely useful vehicle for communication between ecologists. Whilst this situation is likely to continue it is becoming increasingly clear that if we are to use information relating to species to maximum effect two developments are necessary.

(1) Estimations of the ecological amplitude of each species; these will be strongly affected by genetic and phenotypic variation and may change over the geographical range of the species.

(2) Location of each species within a universal classification of organisms into functional types (strategies) (*sensu* Raunkiaer, 1934; Ramensky, 1938; MacArthur & Wilson, 1967; Grime, 1974; Southwood, 1977; Chapin, 1980; Pugh, 1980).

Despite their generality and predictive power, theories bearing on the second of these objectives remain somewhat controversial, partly one suspects because they traverse some of the tribal divisions of biology and more understandably because they are not easily tested by the accepted research tactic of analysing microevolutionary change within contemporary species and populations.

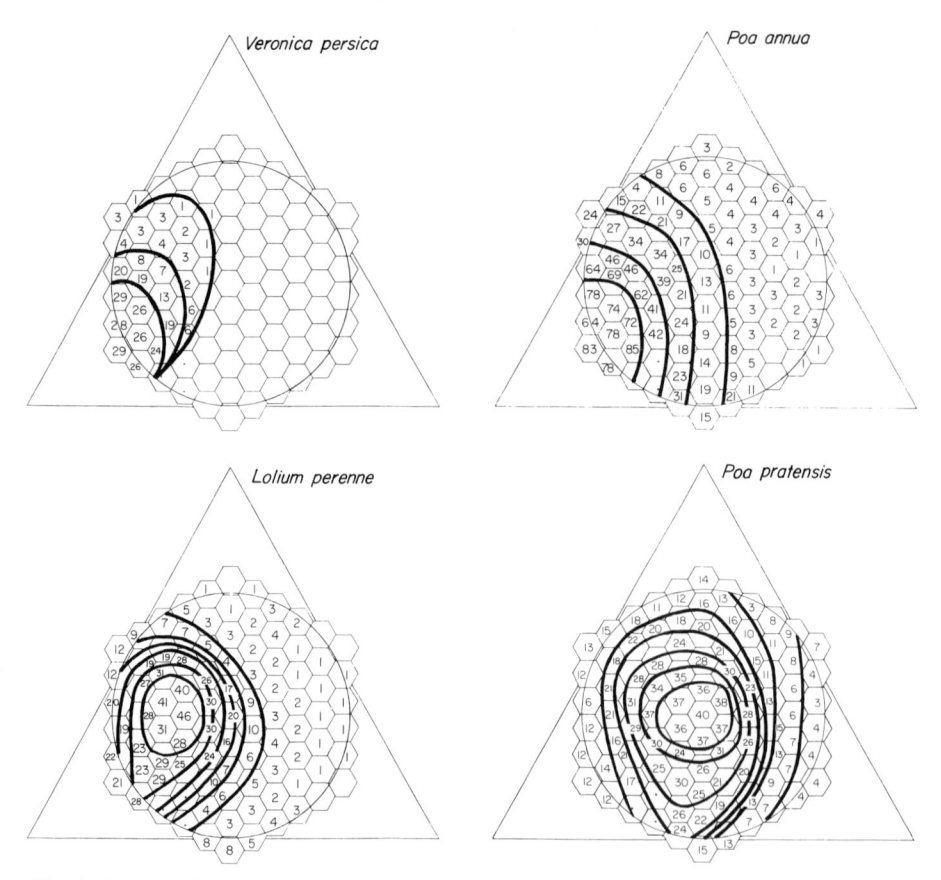

Fig. 1. Contour diagrams describing the frequency of occurrence of eight herbaceous species in a matrix of functionally defined vegetation types. The procedure used in the triangular ordination is described in Appendix 1. Values indicate the percentage of metre-square samples containing

Implicit in attempts to recognize primary ecological strategies is the assertion that organisms share common design constraints and encounter an essentially similar templet of habitat types (Southwood, 1977). The inevitable consequence of this interaction between organism and environment is a limitation of the ecological range of the individual and the channelling of evolution along predictable paths of specialization. In our quest to understand the processes which determine the flora of the British Isles, strategy concepts with their capacity to draw together information on diverse organisms and taxa and from very different research fields provide a possible framework on which to base a functional description of vegetation (Grime, 1984a). This is not to suggest however that strategy theories should be used merely to conduct a sterile 'typing' of British plants. Satisfying insights into the forces which have shaped the flora and continue to affect it can be obtained only by recognizing that the present ecologies of species and populations are strongly conditioned by more ancient patterns of ecological specialization which, as we shall see under the next heading, are themselves susceptible to strategic analysis.

The data in Figure 1 provide an illustration of the use of strategy concepts to describe in broad terms the ecological amplitude of some common British species.

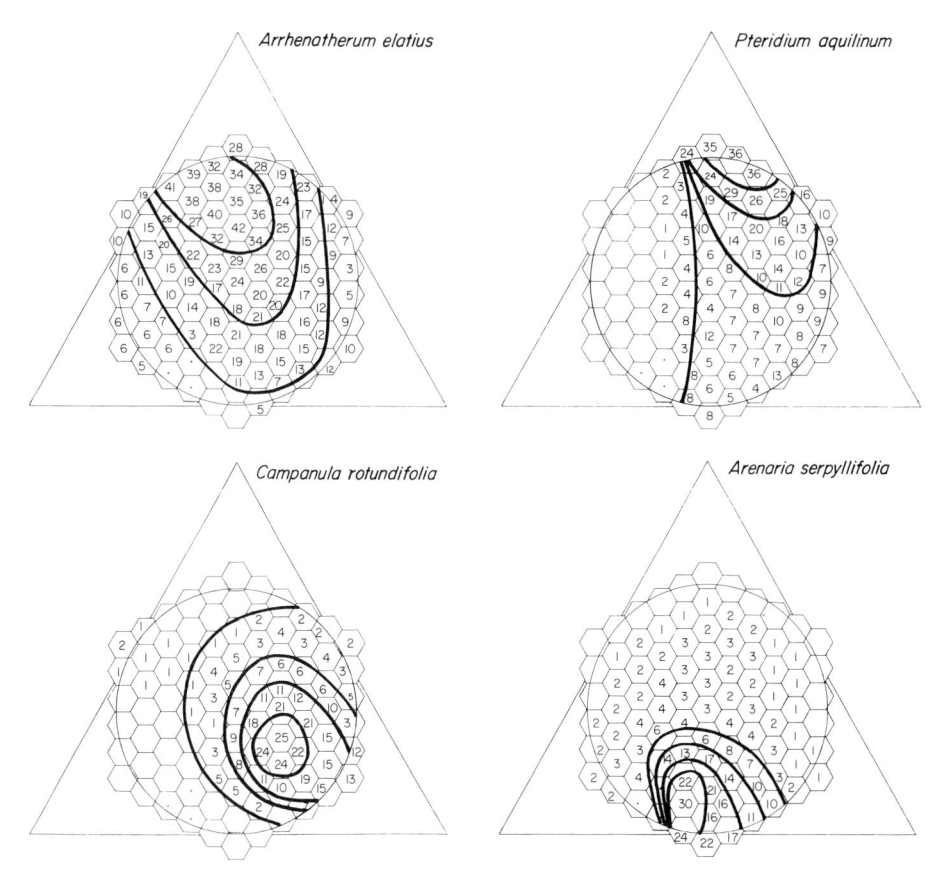

[Figures: Triangular ordination diagrams for *Arrhenatherum elatius*, *Pteridium aquilinum*, *Campanula rotundifolia*, and *Arenaria serpyllifolia*.]

the species. Contours drawn by eye,* insufficient data. Axes of the matrix and positions assigned to marker species are as described in Appendix 1.

The figures are based upon a triangular ordination of 2008 vegetation samples drawn from the major habitats of the Sheffield region. The contours in each diagram plot the frequency of occurrence of the species in a matrix of vegetation types within which each metre-square vegetation sample has been located by reference to selected characteristics of life-history, morphology, phenology and reproduction of the component species. A brief account of the ordination procedure and the strategy concepts upon which it is based are provided in Appendix 1. In order to assist interpretation of the contour diagrams, lists are provided in Appendix 2 of some of the plant characteristics which may be expected to become increasingly prominent in populations, species and vegetation types as we approach the respective corners of the triangle.

When the distributions in Figure 1 are examined many differences are apparent in terms of both the centre and spread of the contours.

In *Veronica persica* there is a very compact distribution. Restriction to the left-hand corner of the diagram indicates that this ephemeral species exploits productive, heavily disturbed vegetation but is unable to colonize infertile habitats and is sensitive to competition from perennial species.

Poa annua displays a similar concentration in the 'ruderal' corner but expands into other parts of the model. This distribution is consistent with results of experimental studies (Law, Bradshaw & Putwain, 1977) which show that in addition to the ephemerals of disturbed ground there are perennial phenotypes of *P. annua* which are particularly common in productive pastures.

The contours for *Lolium perenne* again suggest ruderal characteristics but they are centred rather higher, indicating sensitivity to very heavy disturbance and some ability to persist in perennial communities. The frequency of occurrence of *L. perenne* falls away sharply towards the apex and right-hand side of the triangle, reflecting both the failure of the species to survive in competition with large perennials in productive undisturbed habitats and its exclusion from infertile sites.

The distribution of *Poa pratensis* is focused in the centre of the triangle, suggesting that the species is particularly associated with vegetation experiencing moderate intensities of stress, disturbance and competition, conditions likely to obtain in the pasture habitats exploited by the species. The contours indicate that, to a remarkable extent, *P. pratensis* is able to extend its range into widely different vegetation types. In our present state of knowledge it is uncertain to what extent this wide amplitude is the result of genotypic variation.

A wide strategic range is also suggested by the diagram for *Arrhenatherum elatius*. The highest occurrence coincides with fertile relatively undisturbed conditions (road verges, derelict land, etc.) but the contours also extend towards the base of the triangle, a pattern which may be explained, at least in part, by the occurrence of prostrate genotypes of *A. elatius* shown to be unusually resilient when subjected to defoliation (Mahmoud, Grime & Furness, 1975).

Pteridium aquilinum shows a high frequency of occurrence towards the apex of the triangle, a distribution which suggests that the species is a strong competitor (*sensu* Grime, 1979) with the potential to monopolize plant communities of acidic but relatively fertile undisturbed sites. The complete absence of the species from the left-hand side of the matrix reflects the failure of *P. aquilinum* to exploit frequently disturbed habitats. From the descending contours on the right there is evidence of the ability to persist in some unproductive communities such as those associated with heathlands, and heavily shaded and/or highly acidic woodland herb layers.

The contour pattern of *Campanula rotundifolia* indicates a well-defined ecology. The species appears to be intolerant of high intensities of disturbance and competition and is restricted to relatively unproductive vegetation. *Arenaria serpyllifolia* is concentrated in the area of the triangular model which corresponds to conditions of moderately severe intensities of both stress and disturbance. This accurately portrays the ecology of the species, which is a small winter-annual associated with localities where infertile soils are subjected to disturbance by drought, solifluction or animal activity.

These examples illustrate the potential of strategy concepts to convey in one diagram some of the main features of the ecology of a species. However, despite the large number of plant characteristics which vary in association with the axes of the triangular model (Appendix 2) finer levels of analysis, involving additional plant attributes, are required to explore other dimensions of the ecology of a species. These could include, for example, the use of standardized laboratory measurements of (1) plant response to mineral nutrients and soil toxins (e.g. Jeffries & Willis, 1964; Hackett, 1965; Gigon & Rorison, 1972) to characterize specialization in relation to the 'calcicole–calcifuge' axis, (2) nuclear DNA content (Grime & Mowforth, 1983) as an index of the rate and timing of spring growth and (3) seed morphology and germination characteristics (Noble & Slatyer, 1979; Thompson & Grime, 1979; Grime et al., 1981) to allow recognition of regenerative strategies.

FAMILIES

In Table 2, a survey of the herbaceous vegetation of the Sheffield region (Grime, Hodgson & Hunt, 1985) has been used to examine the affinity of pteridophytes and common families of angiosperms for each of the seven major habitat types present in the region. In order to allow direct comparisons between species differing in abundance within the survey area, calculations of affinity (a) for a particular habitat (H) are based upon the formula:

$$a = \frac{\% \text{ occurrence in } H}{\% \text{ occurrence in all major habitats of the region}}.$$

Table 2. *Estimates of the affinity of pteridophytes and selected families of angiosperms for the seven major types of habitat occurring in the Sheffield region*

	Wetland		Skeletal*		Arable		Grassland		Spoil		Wasteland†		Woodland	
Taxon	−	+	−	+	−	+	−	+	−	+	−	+	−	+
Caryophyllaceae	11	2	8	5	8	5	11	2	4	9	9	4	10	3
Compositae	32	3	11	24‡	22	13	19	16	2	33‡	1	24	32	3‡
Cruciferae	7	4	4	7	8	3	5	6	3	8	4	7	9	2
Cyperaceae	3	13‡	13	3	16	0	9	7	15	1‡	6	10	14	2
Gramineae	45	10	29	26	43	12	23	32‡	27	28	14	41‡	41	14
Juncaceae	3	7‡	7	3	9	1	5	5	8	2‡	3	7	8	2
Labiatae	9	4	8	5	9	4	9	4	6	7	4	9	8	5
Leguminosae	13	1	10	4	9	5	2	12‡	1	13‡	1	13‡	11	3
Onagraceae	3	5	3	5	8	0	8	0‡	2	6	7	1‡	6	2
Polygonaceae	8	4	12	0‡	6	6	9	3	4	8	5	7	10	2
Ranunculaceae	4	7‡	9	2	9	2	7	4	7	4	6	5	9	2
Rosaceae	9	1	6	4	9	1	7	3	6	4	5	5	2	8‡
Scrophulariaceae	10	3	5	8	9	4	9	4	6	7	10	3‡	11	2
Umbelliferae	10	3	9	4	11	2	8	5	7	6	2	11	7	6
Pteridophyta	7	3	4	6	9	1	9	1	7	3	8	2‡	5	5
All common species	229	94	186	137	248	75	190	133	141	182	129	194	239	84

In each habitat the species in each taxon are classified into two groups (−, +) by reference to an index of affinity (see text).
− signifies index < 1·0; + signifies index ⩾ 1·0.
* Rock outcrops, cliffs, screes and walls.
† Derelict land with herbaceous vegetation.
‡ Ratio significantly different ($P < 0.05$) from that for all common species.

The data in Table 2 reveal several families with distinctive ecologies. Particularly clear is the affinity of the Ranunculaceae, Juncaceae and Cyperaceae for wetland and the failure of the last to exploit arable and spoiled land. Polygonaceae are relatively frequent in the disturbed open vegetation of skeletal habitats and arable land, whilst the Compositae are prominent in skeletal habitats and spoil but are relatively unsuccessful in arable and woodland. Predictably we find that the Gramineae and Leguminosae are especially associated with grassland and derelict herbaceous vegetation habitats which are unfavourable to Onagraceae. Of the remaining patterns evident in Table 2 perhaps the most striking is the strong positive association of the Rosaceae with woodland habitats.

It would be a mistake to assume that these patterns reflect in every case avenues of ecological specialization which are characteristic of plant families. The flora of Britain is both recent and depauperate (Godwin, 1975; Walters, 1984) and it is quite clear that in families such as Leguminosae and Compositae the consistent behaviour apparent in Table 2 is merely that of the Temperate European components of the taxa concerned. However, these considerations do not detract from the important observation that major plant families, however incompletely represented in the British flora, exhibit distinctive ecologies within our landscape. This invites re-examination of the relationships between taxonomic characters and ecology (Hodgson, 1984a–c) and suggests that at various stages in the evolution of plants specializations have occurred which continue to exercise surprisingly persistent constraints upon the ecology of contemporary families, tribes, genera and species.

Further research is needed to identify the nature of the ecological specializations which have taken place in particular families. Some, e.g. nitrogen-fixing symbioses and hard-coat seed dormancy in Leguminosae, development of intercalary meristems in Gramineae and production of numerous wind-dispersed seeds in Onagraceae, are already well known, whilst pointers to others are available from large-scale comparative studies of germination and seedling growth in British species (Grime & Hunt, 1975; Grime et al., 1981).

One way in which to examine ecological specialization in plant families is to compare the range of primary strategies exhibited by constituent species. In Figure 2 this has been attempted for some of the commoner families and species of the Sheffield region using again the method of triangular ordination described in Appendix 1. In order to simplify presentation each species is represented by a single dot, the location of which corresponds to the centre of a contour diagram of the type illustrated in Figure 1.

Examination of the distributions in Figure 2 reveals that the common herbaceous plants of the Sheffield region form a complete array within the vegetation matrix. When families are considered separately, however, some distinctive patterns are observed. A concentration of species in the area of the model corresponding to disturbed fertile habitats is apparent in the Caryophyllaceae, Cruciferae and Polygonaceae. In marked contrast the Cyperaceae and Rosaceae are more common in relatively undisturbed conditions, and they include a high proportion of species which are confined to unproductive habitats. It is interesting to note, however, that four of the most prominent contributors to the contemporary British flora in terms of both number of species and biomass (Compositae, Gramineae, Labiatae and Leguminosae) exhibit wide-ranging distributions.

Also depicted in Figure 2 is the distribution of the commoner pteridophytes of the Sheffield region. These occur mainly in relatively undisturbed, unproductive vegetation and show little penetration into heavily disturbed environments. An examination of the features which limit the ecological range of pteridophytes has been presented elsewhere (Grime, 1984b). The main conclusions are that the contribution of pteridophytes to the flora of the Sheffield region is restricted by the absence of ephemeral and vernal life-forms, the scarcity of shoot systems resilient under defoliation, the low relative growth-rates of many species and the susceptibility of gametophytes and young sporophytes to competition from

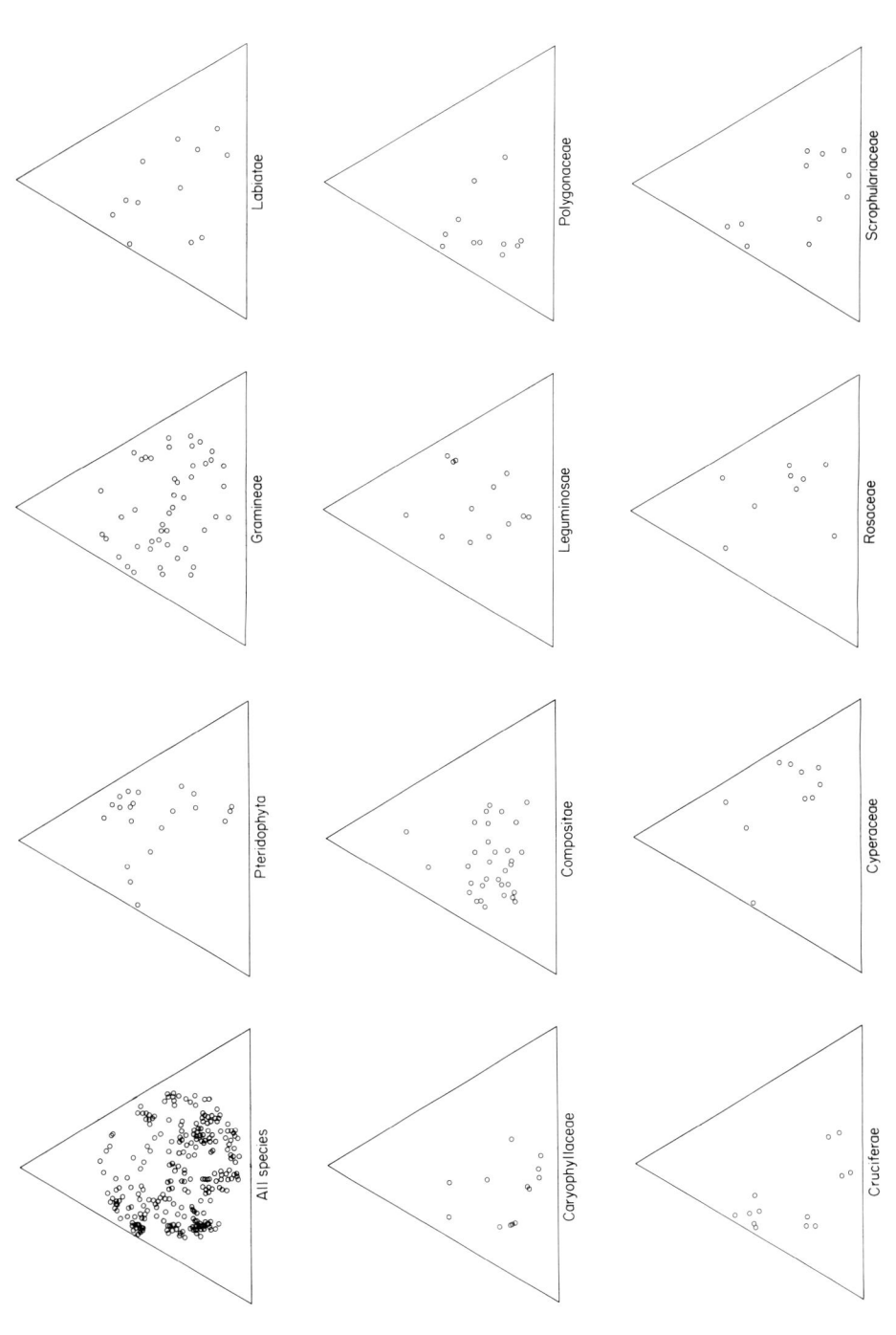

Fig. 2. The distribution of contour centres of common species of the Sheffield region classified by taxon. Each circle indicates the centre of the contour diagram for a species (see Fig. 1). Axes of the matrix and positions assigned to marker species are as described in Appendix 1.

Table 3. *Examples of community properties which may be predicted from characteristics of the most abundant component species*

Characteristics of most abundant species	Community properties	Examples
Annuals with high growth rates and producing seeds with a high level of innate dormancy	High resilience under severe disturbance	Garden weeds, drift-line assemblages
Long-lived, slow-growing, evergreen perennials	Low resilience under severe disturbance	Woodland herb layers such as those containing *Deschampsia flexuosa* or *Sanicula europaea*
Large perennials with high growth rates and potential for extensive lateral spread	Competitive dominance and low species diversity	Nettle (*Urtica dioica*) patches, derelict roadsides with *Arrhenatherum elatius*
Long-lived, slow-growing perennials producing large numbers of small widely dispersed propagules	Restriction to 'open' unproductive and often relatively inaccessible sites	Bryophyte and fern communities of cliffs, walls and tree trunks
Perennials with high growth rates and producing large numbers of widely dispersed propagules	Capacity to recur widely in landscapes subject to spatially unpredictable habitat disturbance	Patches of *Chamerion angustifolium* in disturbed forest and urban land. Stands of *Epilobium hirsutum* in disturbed eutrophic mire
Large interspecific differences in nuclear DNA content	Strongly developed temporal niche-differentiation	*Lolium perenne*–*Trifolium repens*, *Festuca rubra*–*Agrostis capillaris* and *Bromus erectus*–*Brachypodium pinnatum* grasslands. *Quercus petraea*–*Hyacinthoides non-scripta* woodland

herbaceous angiosperms. This interpretation leads inescapably to the prediction that with the notable exceptions of *Pteridium aquilinum* and certain species of *Equisetum*, pteridophytes may be expected to decline under the impact of current changes in land-use which essentially involve disturbance and eutrophication.

It has been proposed (Hodgson, 1984a–c) that similar arguments may be used to explain differences between families of angiosperms in their abundance and response to changing patterns of land-use in Britain. Hence we may suspect that the current success (measured in terms of species and biomass) of Gramineae and Compositae may be traced to ancestral characteristics of these families which have been conducive to the development of ruderal and competitive strategies (*sensu* Grime, 1979) both of which are favoured by recent changes in land-use. Conversely, it seems likely that the predominantly stress-tolerant characteristics of the majority of Cyperaceae and Rosaceae will relegate these families to a declining role in the future countryside of Britain.

COMMUNITIES

If we again return to the quotation in the Introduction to this paper it is interesting to see that Clapham's approach to the plant community is to attempt to understand it as a function of the characteristics of its component species. This is such a logical proposal that it is perhaps surprising to find that it has failed to attract general support. Undoubtedly, a key factor which has limited progress towards the objective of identifying the functional characteristics of plant communities in Britain has been the slow progress in documenting in a standardized way the distinctive features of common plants. I shall not attempt to analyse the reasons for this failure, although it is relevant to point out that it coincides with the narrowing focus and increasing fragmentation of much ecological work.

One of the penalties arising from our woefully incomplete knowledge of the characteristics of species has been the inadequacy of research into the properties of the plant community such as resistance, resilience, species diversity and the rate and trajectory of successional change, all of which are of vital concern to vegetation management. In the absence of the basic information needed to apply a predictive and experimental approach, these problems have tended to remain becalmed in the Doldrums of mathematical ecology.

Where sufficient knowledge of the characteristics of species is available some progress has been made in prediction and interpretation of community properties (Nobel & Slatyer, 1979; Shepherd, 1981; Buttenschon & Buttenschon, 1982; Leps, Osbornova-Kosinova & Rejmanek, 1982) and, in Table 3, examples are provided of the application of this approach to some plant communities of common occurrence in Britain.

Perhaps the most searching test of the value of the reductionist, species-centred approach to plant communities of the British Isles is to use it in attempts to analyse the mechanisms which allow large numbers of species to co-exist in the ancient species-rich grasslands of chalk and limestone. Figure 3 provides an illustration of how such an analysis might commence by comparing the species composition of two metre-square samples of rather different types of species-rich grassland both in North Derbyshire and both situated within the 350 to 750 g m^{-2} corridor in seasonal peak of shoot biomass+litter known to permit high diversity in herbaceous vegetation (Al-Mufti *et al.*, 1977). Each community is represented by a triangular diagram in which the abundance of each species is indicated by the

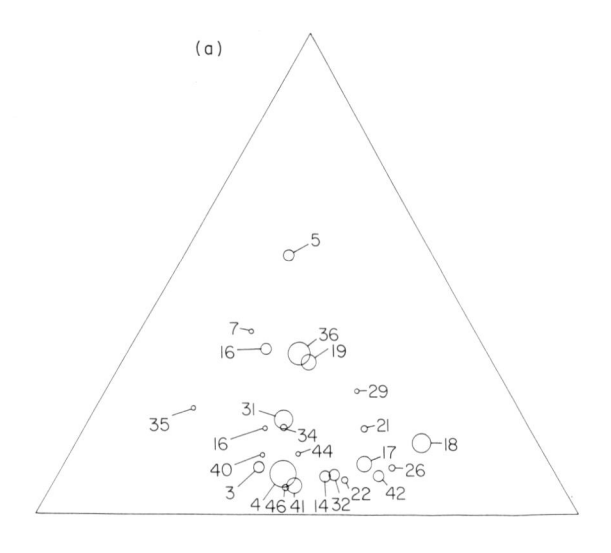

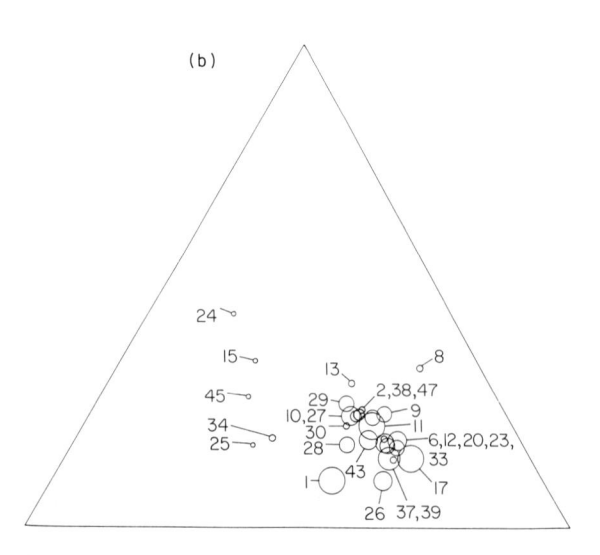

Fig. 3. Comparison of the 'strategic composition' of two contrasted types of species-rich calcareous grassland. Each diagram refers to a metre-square sample of vegetation. Component species are located by reference to the contour centres (see Fig. 1). Abundance of the species in the vegetation sample is indicated by the size of the circle. (a) South-facing grassland on shallow soil over fissured limestone. (b) North-facing grassland in seepage zone. Key to species: 1, *Anemone nemorosa*; 2, *Anthoxanthum odoratum*; 3, *Aphanes arvensis*; 4, *Arenaria serpyllifolia*; 5, *Arrhenatherum elatius*; 6, *Avenula pratensis*; 7, *Bellis perennis*; 8, *Brachypodium sylvaticum*; 9, *Briza media*; 10, *Campanula rotundifolia*; 11, *Carex caryophyllea*, 12, *Carex panicea*; 13, *Centaurea nigra*; 14 *Cerastium semidecandrum*; 15, *Dactylis glomerata*; 16, *Desmazeria rigidum*; 17, *Erophila verna*; 18, *Festuca ovina*; 19, *Festuca rubra*; 20, *Galium sterneri*; 21, *Galium verum*; 22, *Geranium molle*; 23, *Helianthemum nummularium*; 24, *Heracleum sphondylium*; 25, *Hypericum perforatum*; 26, *Koeleria macrantha*; 27, *Leontodon hispidus*; 28, *Linum catharticum*; 29, *Lotus corniculatus*; 30, *Luzula campestris*; 31, *Medicago lupulina*; 32, *Myosotis ramosissima*,; 33, *Pimpinella saxifraga*; 34, *Plantago lanceolata*; 35, *Poa annua*; 36, *Poa pratensis*; 37, *Polygala vulgaris*; 38, *Potentilla erecta*; 39, *Sanguisorba minor*; 40, *Sagina apetala*; 41, *Saxifraga tridactylites*; 42, *Sedum acre*; 43, *Succisa pratensis*; 44, *Trifolium dubium*; 45, *Trifolium pratense*; 46, *Veronica arvensis*; 47, *Viola riviniana*.

Ecology of the contemporary British Flora

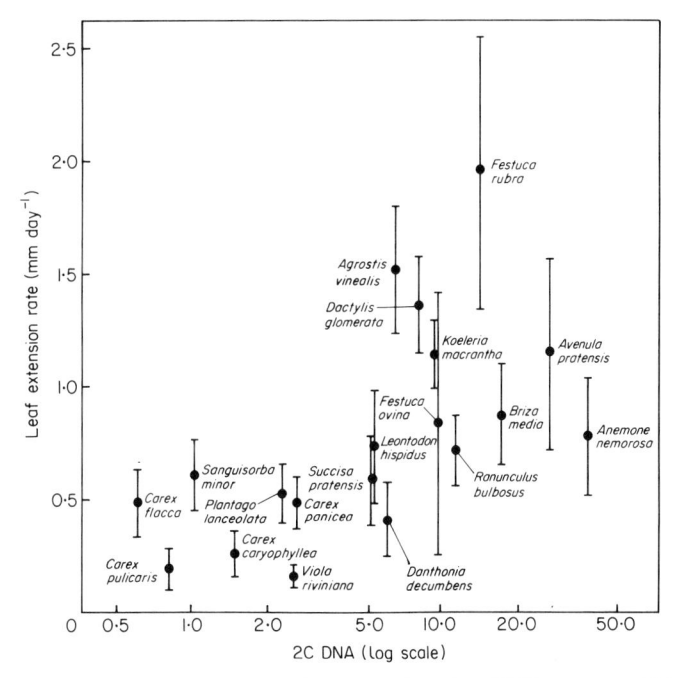

Fig. 4. Relationship between leaf extension rate and nuclear DNA content within a community of grassland plants. Measurements of nuclear DNA content by Feulgen staining and microdensitometry on root tips. Estimations of leaf extension rate refer to measurements between 24 March and 8 April 1983 on immature leaves of randomly selected shoots in turf transplanted to an experimental garden. The vertical bars indicate 95% confidence limits.

size of a circle, the position of which corresponds to the centre of a contour diagram constructed using the procedure described in Appendix 1.

Reference to the diagram for Community (a) which is situated on a grazed limestone outcrop of fissured limestone on a south-facing slope reveals that the vegetation contains plants of widely contrasted stature, life-history and regenerative biology. The range includes at its extremities potentially robust and fast-growing perennials (*Arrhenatherum elatius*, *Festuca rubra* and *Dactylis glomerata*) and small winter annuals (*Erophila verna*, *Saxifraga tridactylites*, *Veronica arvensis*) suggesting that species-richness in this community is related at least in part to the presence of microhabitats differing in soil depth, mineral nutrient supply and exposure to summer drought.

In marked contrast, Community (b), situated in a north-facing pasture and occupying a seepage zone, is more homogeneous, the majority of the components being small slow-growing perennials forming a compact distribution within the triangle. We are thus alerted to the strong possibility that the mechanisms permitting species co-existence in Community (b) are more subtle than those of Community (a). However, the range of nuclear DNA contents in Community (b) is exceedingly wide (2C DNA 0·6 to 38·5 pg) suggesting that temporal differences in growth may be important in limiting niche overlap. Measurements of leaf growth during the spring (Fig. 4) support this hypothesis.

These two examples describe the preliminary incisions of the dissection necessary to expose the workings of a plant community. Only the patient accumulation of comparative autecological data can provide the instruments required to complete the task.

CONCLUSIONS

Over the last three decades research on the British flora has not advanced on the broad front envisaged by Clapham (1956) and anticipated as a logical development following the appearance of the *Flora of the British Isles* (Clapham, Tutin & Warburg, 1952) and the *Atlas of the British Flora* (Perring & Walters, 1962). Research effort has become more detailed and specialized and the dynamic approach to plant populations presaged in the work of E. J. Salisbury, A. S. Watt and C. O. Tamm has come into its own, fortified by modern techniques of data analysis and computer modelling. These are welcome developments which, in particular, have added temporal and genetic components to our thinking about vegetation. It is important, however, that we should not be seduced by the precision and elegance of current techniques into an excessive preoccupation with the detailed functioning of contemporary populations of amenable species. If the emergence of plant ecology from its descriptive phase is not to be illusory, the subject must also include forms of research which are broadly based and capable of placing in general perspective the selection forces and design constraints which have interacted in the past as well as the present to determine the current ecology of populations, species, families and communities.

ACKNOWLEDGEMENTS

I wish to thank all members of the Unit of Comparative Plant Ecology who participated in the field and laboratory studies drawn upon in this paper. In particular I am indebted to Dr J. G. Hodgson for the opportunity to refer to unpublished manuscripts. This paper is based upon research supported by the Natural Environment Research Council.

REFERENCES

AL-MUFTI, M. M., SYDES, C. L., FURNESS, S. B., GRIME, J. P. & BAND, S. R. (1977). A quantitative analysis of shoot phenology and dominance in herbaceous vegetation. *Journal of Ecology*, 65, 759–791.
BUTTENSCHON, J. & BUTTENSCHON, R. M. (1982). Grazing experiments with cattle and sheep on nutrient poor, acidic grassland and heath. I. Vegetation development. *Natura Jutlandica*, 21, 1–48.
CHAPIN, F. S. (1980). The mineral nutrition of wild plants. *Annual Review of Ecology and Systematics*, 11, 233–260.
CLAPHAM, A. R., TUTIN, T. G. & WARBURG, E. F. (1952). *Flora of the British Isles*, 1st Edn. Cambridge University Press, Cambridge.
CLAPHAM, A. R. (1956). Autecological studies and the 'Biological Flora of the British Isles'. *Journal of Ecology*, 44, 1–11.
GIGON, A. & RORISON, I. H. (1972). The response of some ecologically distinct plant species to nitrate-and to ammonium-nitrogen. *Journal of Ecology*, 60, 93–102.
GODWIN, H. (1975). *The History of the British Flora. A Factual Basis for Phytogeography*, 2nd Edn. Cambridge University Press, Cambridge.
GRIME, J. P. (1974). Vegetation classification by reference to strategies. *Nature*, 250, 26–31.
GRIME, J. P. (1979). *Plant Strategies and Vegetation Processes*. J. Wiley & Sons, Chichester.
GRIME, J. P. (1983). Prediction of weed and crop response to climate based upon measurements of nuclear DNA content. In: *Aspects of Applied Biology*, vol. 4 (Ed. by J. C. Caseley), pp. 87–98.
GRIME, J. P. (1984a). Towards a functional description of vegetation. In: *Population Structure of Vegetation* (Ed. by J. White & J. Beeftink), Dr Junk, The Hague. (In press).
GRIME, J. P. (1984b). Factors limiting the contribution of pteridophytes to a local flora. In: *The Biology of Pteridophytes* (Ed. by A. F. Dyer & C. N. Page). Edinburgh University Press. (In press).

GRIME, J. P., HODGSON, J. G. & HUNT, R. (1985). *Comparative Plant Ecology. A Functional Approach to Common British Species and Communities.* George Allen & Unwin Ltd, London. (In press).
GRIME, J. P. & HUNT, R. (1975). Relative growth-rate: its range and adaptive significance in a local flora. *Journal of Ecology*, **63**, 393–422.
GRIME, J. P., MASON,, G., CURTIS, A. V., RODMAN, J., BAND, S. R., MOWFORTH, M. A. G., NEAL, A. M. & SHAW, S. (1981). A comparative study of germination characteristics in a local flora. *Journal of Ecology*, **69**, 1017–1060.
GRIME, J. P. & MOWFORTH, M. A. (1983). Variation in genome size – an ecological interpretation. *Nature*, **299**, 153–155.
HACKETT, C. (1965). Ecological aspects of the nutrition of *Deschampsia flexuosa* (L.) Trin. II. The effects of Al, Ca, Fe, K, Mn, N, P, and pH on the growth of seedlings and established plants. *Journal of Ecology*, **53**, 315–333.
HARPER, J. L. (1982). After description. In: *The Plant Community as a Working Mechanism* (Ed. by E. I. Newman), pp. 11–25. Special publication No. 1, British Ecological Society. Blackwell Scientific Publications, Oxford.
HODGSON, J. G. (1984a). Commonness and rarity in the Sheffield flora. I. The identity, distribution and habitat characteristics of the common and rare species. *Biological Conservation*, (in press).
HODGSON, J. G. (1984b). Commonness and rarity in the Sheffield flora. II. The relative importance of various environmental determinants of present-day commonness and rarity. *Biological Conservation* (in press).
HODGSON, J. G. (1984c). Commonness and rarity in the Sheffield flora. III. Taxonomic and evolutionary aspects. *Biological Conservation* (in press).
JEFFRIES, R. L. & WILLIS, A. J. (1964). Studies on the calcicole–calcifuge habit. II. The influence of calcium on the growth and establishment of four species in soil and sand cultures. *Journal of Ecology*, **52**, 691–707.
LAW, R., BRADSHAW, A. D. & PUTWAIN, P. D. (1977). Life history variation in *Poa annua*. *Evolution*, **31**, 233–246.
LEPS, J. J., OSBORNOVA-KOSINOVA & REJMANEK, K. (1982). Community stability, complexity and species life-history strategies. *Vegetatio*, **50**, 53–63.
MACARTHUR, R. H. & WILSON, E. D. (1967). *The Theory of Island Biogeography.* Princeton University Press, Princeton, NJ.
MAHMOUD, A., GRIME, J. P. & FURNESS, S. R. (1975). Polymorphism in *Arrhenatherum elatius* (L.) Beauv. ex J. and C. Presl. *The New Phytologist*, **75**, 269–276.
NOBEL, I. R. & SLATYER, R. O. (1979). The use of vital attributes to predict successional changes in plant communities subject to recurrent disturbances. *Vegetatio*, **43**, 5–21.
PERRING, F. H. & WALTERS, S. M. (1962). *Atlas of the British Flora.* Nelson, Edinburgh.
PUGH, G. J. F. (1980). Strategies in fungal ecology. *Transactions of the British Mycological Society*, **75**, 1–14.
RAMENSKY, L. G. (1938). *Introduction to the Geobotanical Study of Complex Vegetations.* Selkozgiz, Moscow.
RAUNKIAER, C. (1934). *The Life Forms of Plants and Statistical Plant Geography*; being the collected papers of C. Raunkiaer, translated into English by H. G. Carter, A. G. Tansley and Miss Fansball. Clarendon Press, Oxford.
SHEPHERD, S. A. (1981). Ecological strategies in a deep water red algal community. *Botanica Marina*, **XXIV**, 457–463.
SOUTHWOOD, T. R. E. (1977). Habitat, the templet for ecological strategies? *Journal of Animal Ecology*, **46**, 337–365.
THOMPSON, K. & GRIME, J. P. (1979). Seasonal variation in the seed banks of herbaceous species in ten contrasting habitats. *Journal of Ecology*, **67**, 893–922.
WALTERS, S. M. (1984). The relationship between British and European floras. In: *The Flora and Vegetation of Britain.* A symposium organized by the New Phytologist Trust and the University of Sheffield to honour the 80th birthday of Professor A. R. Clapham. University of Sheffield. 19th May 1984.

Appendix 1.

Brief description of the triangular ordination used in Figures 1 to 4

Theory

The ordination procedure rests upon the assertion (Ramensky, 1938; Grime, 1974) that the primary mechanism controlling the nature and distribution of plant populations, species and communities arises from three selection processes which operate in the present and have also exerted a dominant influence in the evolution of plants. Two of these selection processes may be defined as *stress*, consisting of the external constraints on dry-matter production, and *disturbance*, identified as the destruction of biomass. At high intensities both stress and disturbance select for particular types of life-history and physiology (Appendix 2). Where the intensities of stress and disturbance are low, rapid rates of resource capture and growth are possible and a third selection process (competition for light, water, mineral nutrients and space) becomes of overriding importance and selects for a quite different set of plant characteristics (Appendix 2). In addition to the strategies associated with high intensities of either stress, disturbance or competition there are others which exploit the various intermediate conditions, corresponding to particular equilibria between the three selection forces; these may be described by means of a triangular model (Fig. 5). A more detailed account of these concepts has been provided elsewhere (Grime, 1979).

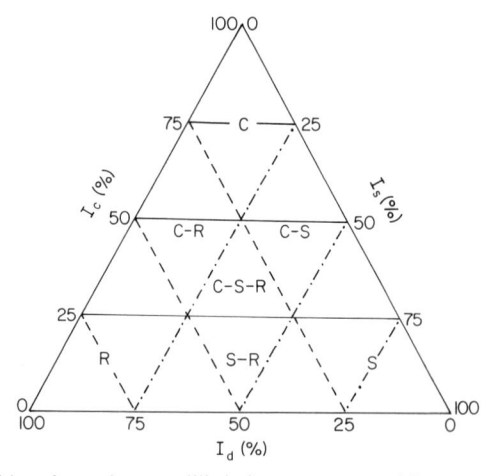

Fig. 5. Model describing the various equilibria between competition, stress and disturbance in vegetation and the location of primary and secondary strategies. C, competitor; S, stress-tolerator; R, ruderal; C-R, competitive ruderal; S-R, stress-tolerant ruderal; C-S, stress-tolerant competitor; C-S-R strategist, I_c, relative importance of competition (———); I_s, relative importance of stress (—·—·—); I_d, relative importance of disturbance (———).

Procedure

The initial step in the ordination was to attempt to classify with respect to strategy the herbaceous species of the Sheffield region. The method involved the use of a dichotomous key (Fig. 6) based upon characteristics of life-history, morphology and phenology and allowed species to be assigned to one of seven positions (Fig. 6, inset) corresponding to a characteristic pair of stress and disturbance coordinates and situated within a circular area of the triangular model conforming to the strategic range proposed for herbaceous plants (Grime, 1979, fig. 18, p. 73). It should be emphasized that this was merely an approximate and provisional classification which could be applied with certainty to only a restricted number of species. No attempt was made to classify species for which critical data were lacking, and species known to exhibit major variation in life-history and morphology were also omitted. This procedure allowed the classification of 204 species (henceforward described as marker species) including many of the commoner herbaceous plants of the region.

The next step was to ordinate vegetation samples drawn from the range of habitats represented in the Sheffield region. Each 1 m² sample was located in the

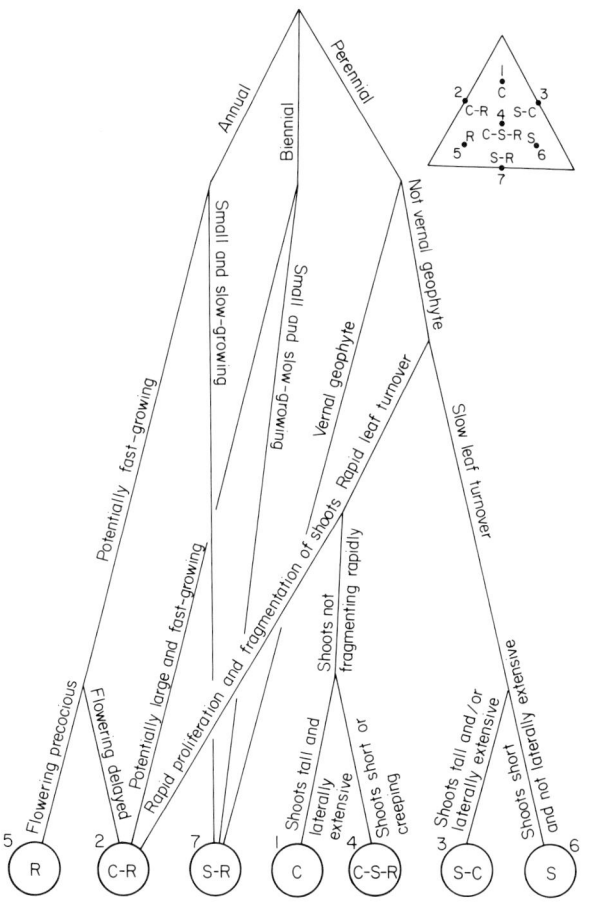

Fig. 6. A dichotomous key to strategies in herbaceous plants used to classify marker species in the triangular ordination. C, competitor; S, stress-tolerator; R, ruderal; C-R, competitive-ruderal; S-R, stress-tolerant ruderal; C-S, stress-tolerant competitor; C-S-R, 'C-S-R strategist'. The location of the strategies within the triangular model (see Figure 5 for axes) is shown by the inset.

triangle by reference to the positions and frequencies of the component marker species. This was achieved by calculating mean stress and disturbance coordinates in which the contribution of each marker species was weighted according to its frequency in the vegetation sample. By this procedure, 2008 vegetation samples were ordinated and found to be distributed fairly evenly within the central circular area of the triangle. This allowed a calculation of the percentage occurrence of each species within each of 91 hexagonal zones of the circle, and this in turn permitted contours to be drawn describing the distribution of each species.

Appendix 2. *Some characteristics of competitive, stress-tolerant and ruderal plants*

	Competitive	Stress-tolerant	Ruderal
(i) Morphology			
1 Life forms	Herbs, shrubs and trees	Lichens, bryophytes, herbs, shrubs and trees	Herbs, bryophytes
2 Morphology of shoot	High dense canopy of leaves. Extensive lateral spread above and below ground	Extremely wide range of growth forms	Small stature, limited lateral spread
3 Leaf form	Robust, often mesomorphic	Often small or leathery, or needle-like	Various, often mesomorphic
4 Canopy structure	Rapidly elevating monolayer	Often multilayered. If monolayer not rapidly elevating	Various
(ii) Life history			
5 Longevity of established phase	Long or relatively short	Long–very long	Very short
6 Longevity of leaves and roots	Relatively short	Long	Short
7 Leaf phenology	Well-defined peaks of leaf production coinciding with periods of maximum potential productivity	Evergreens, with various patterns of leaf production	Short phase of leaf production in period of high potential productivity
8 Phenology of flowering	Flowers produced after (or, more rarely, before) periods of maximum potential productivity	No general relationship between time of flowering and season	Flowers produced early in the life history
9 Frequency of flowering	Established plants usually flower each year	Intermittent flowering over a long life history	High frequency of flowering
10 Proportion of annual production devoted to seeds	Small	Small	Large

		Dormant buds and seeds	Stress-tolerant leaves and roots	Dormant seeds
11	Perennation			
12	Regenerative* strategies	V, S, W, B$_S$	V, W, B$_{SG}$	S, W, B$_S$

(iii) Physiology

13	Maximum potential relative growth rate	Rapid	Slow	Rapid
14	Response to stress	Rapid morphogenetic responses (root–shoot ratio, leaf area, root surface area) maximizing vegetative growth	Morphogenetic responses slow and small in magnitude	Rapid curtailment of vegetative growth, diversion of resources into flowering
15	Photosynthesis and uptake of mineral nutrients	Strongly seasonal, coinciding with long continuous period of vegetative growth	Opportunistic, often uncoupled from vegetative growth	Opportunistic, coinciding with vegetative growth
16	Acclimation of photosynthesis, mineral nutrition and tissue hardiness to seasonal change in temperature, light and moisture supply	Weakly developed	Strongly developed	Weakly developed
17	Storage of photosynthate and mineral nutrients	Most photosynthate and mineral nutrients are rapidly incorporated into vegetative structure but a proportion is stored and forms the capital for expansion of growth in the following growing season	Storage systems in leaves, stems and/or roots	Confined to seeds

(iv) Miscellaneous

18	Litter	Copious, often persistent	Sparse, sometimes persistent	Sparse not usually persistent
19	Palatability to unspecialized herbivores	Various	Low	Various, often high
20	Genome size	Usually small	Various	Small–very small

* Key to regenerative strategies, V, vegetative expansion; S, seasonal regeneration in vegetation gaps; W, numerous small wind-dispersed seeds or spores; B$_S$, persistent seed bank; B$_{SG}$, persistent seedling bank.

NEWFERRY AND THE BOREAL-ATLANTIC TRANSITION

By A. G. SMITH

Department of Plant Science, University College, Cardiff

Summary

The establishment and expansion of alder [*Alnus glutinosa* (L.) Gaertner] in Britain between approx. 8000 and 5000 years ago is discussed in relation to changes of climate and sea-level, and to human influences.

Detailed pollen analyses of Mesolithic layers at Newferry, Co. Antrim are presented. Evidence from this and other sites suggest widespread disturbance of forest cover which, by locally reducing competition, (cf. McVean, 1956) appears to have been one of the factors in the establishment of alder.

Estuarine habitats are suggested as having importance in the spread of alder but the onset of climatic wetness was probably not a key factor (as has often hitherto been supposed) other than in the sense of being permissive.

Key words: Newferry, Boreal-Atlantic transition, alder, pollen analysis, radiocarbon dating.

Introductory Review

This paper sets out to re-examine some of the pollen analytic and stratigraphic evidence for vegetational and environmental change at the classical horizon known as the Boreal-Atlantic transition. The use of these terms has been abandoned by many recent authors, and for quite good reasons. Nevertheless, they still serve as a convenient framework in which to enclose the rise of the pollen curve of alder [*Alnus glutinosa* (L.) Gaertner] which is a prominent feature of all Flandrian pollen diagrams of sufficient age.

The names 'Boreal' and 'Atlantic' are derived from the work of the famous Scandinavian bog geologists, Axel Blytt and Rutger Sernander (for references and some summary in English see Fries, 1965a, b). Working in late Victorian and Edwardian times they interpreted the presence of trees in peat bogs to imply a dry climate in the period they called 'Boreal' and the absence of trees in the superincumbent layer to indicate a wet climate: the 'Atlantic' period. (As is well known, the sequence was considered to have been repeated, giving names, again with climatic implications to the Sub-boreal and Sub-atlantic periods.) Even in 1909, however, we find another Scandinavian worker, Gunnar Andersson putting considerable obstacles in the way of the simple interpretation of tree stump layers in peat as indicative of climatic dryness.

Despite objections such as these the Blytt-Sernander terms, together with their climatic implications, became transferred to the pollen zones that were subsequently defined both on the continent and in Britain. Andersson's doubts have been raised again, however, by the most recent study of tree-layers in peat in Britain. Birks (1975) concludes that little regional climatic significance can be assumed from the occurrence of pine stumps in Scottish bogs, and that they cannot be taken as evidence in support of dry Boreal and Sub-boreal periods. Despite this, there is evidence of Boreal dryness from other sources (see below).

In the 1930's and early 1940's when Godwin was establishing his pollen zones for England he concluded that his pollen zones V and VI corresponded to the Boreal period (1940a, b, p. 245). The transition between his pollen zones VI and VII was characterized by a sudden replacement of pine by alder as the most abundant tree pollen. In 1940 he expressed the view that the sudden explosive development of alder over an enormously wide area could hardly be due to anything other than a great increase in wetness. In writing his great book *The History of the British Flora* Godwin largely adhered to this interpretation (Godwin 1975, p. 464). He stated that the expansion of alder at the opening of pollen zone VIIA was in part a thermal response but that the 'sudden massive rise in the alder pollen curve can only be the response to general waterlogging of lowland plains and valleys...'

A similar view of the expansion of alder in the Flandrian is taken by West (1980). He speaks of a general increase in damp habitats at this time following the increased oceanicity of the climate which resulted in the spread of ombrogenous mires in the west and north of Britain. In considering the temperate stages of the Pleistocene in East Anglia he notes variable, but generally low, pollen values of *Alnus* in the Ipswichian Interglacial which he has considered (West, 1961) as probably having a relatively continental climate.

The spread of upland peats in Atlantic times in northern Britain has been mentioned by a number of authors as possible evidence of increasing wetness. The date of upland peat initiation has been shown to be variable, however, both in the Pennines (Tallis, 1964) and in Wales (Chambers, 1981). Tallis considered that the dates for ombrogenous peat initiation available in 1964 showed a latitudinal effect with peat beginning to form first in northern Britain. With the increase in the number of sites investigated in recent years and increasing evidence of the role of man in blanket peat initiation (e.g. Moore, 1975; Smith, 1975; Smith *et al.*, 1981) this is a topic which demands an up-to-date review. We shall allude later to the relationship of Mesolithic activity to blanket peat initiation.

One of the more compelling arguments for regarding the Boreal-Atlantic transition as involving an increase of wetness is a comparison with the later Boreal period in which there is substantial evidence for a period of dryness. This evidence covers a wide area. It comes for instance from the East Anglian Breckland (Hockam Mere: Godwin & Tallentire, 1951), East Yorkshire (Brandesburton: Clark & Godwin, 1956), the English Lake District (Pennington, 1964, 1970, 1981), Cornwall (Brown, 1977), Ireland as a whole (Jessen, 1949), and, more specifically, from County Down (Singh & Smith, 1973).

Before it was possible to obtain a good independent chronology for post-glacial history, as it now is by means of radiocarbon dating, there was nothing that denied the probable synchroneity of the Boreal-Atlantic transition. Indeed, because great emphasis was placed on the primacy of climatic control (e.g. Godwin, 1940a, p. 372) and because the Boreal-Atlantic transition appeared to conform to the scheme of climate change proposed by Blytt & Sernander, it was probably quite generally thought of as synchronous.

In a review of the impact of radiocarbon dates on the dating of pollen zones published in 1973 (Smith & Pilcher) it became apparent, however, that the rise of the *Alnus* curve – the major feature of the transition – is strongly time transgressive. The 'rational limit' for *Alnus*; that is, the point at which the curve begins to rise to substantial values was shown to vary in age from before 7500 BP to just before 5000 BP. The range is thus some 2500 years. Taking a different definition of the Boreal-Atlantic transition – that employed by Mitchell (1951) – the crossing of a falling *Pinus* curve with a rising *Alnus* curve, the range was shown to be even greater. This conclusion may be developed by reference to the recently published isopoll maps of Huntley & Birks (1983). These show no values of *Alnus* in Britain exceeding 2% at 8500 BP. The 2% level is taken by the authors to represent normally a local but sparse presence. Their data suggest, then, that alder had not immigrated to Britain by 8500 BP. By 8000 BP, however, the 2% value had been exceeded at one site in south-east England where immigration might logically have been expected to begin. The authors interpret their maps as showing that alder reached north-west Scotland and eastern Ireland by 6500 BP.

The area between the 2 and 10% isopolls of Huntley & Birks might be taken as representing the migrating front in which alder was assuming some importance in the local vegetation. While the maps are not perhaps intended to be interpreted in such detail they appear to suggest a wave of alder expansion fanning out northwards and westwards from south-east England after 8000 BP. The zone of expansion appears (in broad terms) to have reached the Welsh borders by 7500 BP; into the Scottish borders by 7000 BP; into southern Scotland and eastern Ireland by 6500 BP; central Ireland and north-west Scotland by 6000 BP; and towards western Ireland by 5000 BP. This picture is broken, by an early outpost of alder in Cumbria by 8000 BP though, at the time of writing, the author has been unable to trace any published radiocarbon evidence of this. The oldest radiocarbon date for the rational limit of *Alnus* in Cumbria appears to be 7560 ± 160 BP (Y-2361) (Burnmoor Tarn; Pennington, 1970).

The non-synchroneity in Britain of the expansion of alder pointed out by Smith & Pilcher (1973) is confirmed by these studies but doubtless the broad picture will be modified as data accumulate. The extent of the diachroneity is further illustrated in Figure 1 in which the ^{14}C age of the rational limit (Smith & Pilcher, 1973, p. 904) of the *Alnus* curve is plotted for a range of sites. The rational limit, while somewhat subjective, may perhaps be taken as the point at which alder began to become substantially established in an area.

The fact that the expansion of *Alnus* is not synchronous does not itself negate a climatic interpretation. It could be argued that in the areas where the expansion was late some critical threshhold was not crossed at the time of the earlier expansion in other areas. The difficulty with that explanation is that the areas in which the expansion of alder was late appear to be northern and western, and often upland. In any climatic shift towards wetness it would surely have been in just these oceanic areas that the effects would have been first seen.

The late rise of the alder curve in the north and west of the British Isles could then primarily reflect alder's natural migration. Indeed, such a broad conclusion is drawn by Huntley & Birks. They deduce migration rates for alder in Europe of between 0.5 and 2 km year^{-1}, but decreasing with the passage of time. They mention the lack of suitable nutrient-rich soils, and competition, as limiting factors in the spread of alder. Climatic change, however, they do not hold responsible for controlling the rate and direction of migration. The general lateness of the immigration of alder into western Europe Huntley & Birks attribute largely to the distance of its glacial refugia. It is of interest to note, nevertheless, that they do see climatic change as possibly playing some role locally; for instance in western Scotland.

Godwin (e.g. 1975, p. 464, 466) has concluded that alder must have been

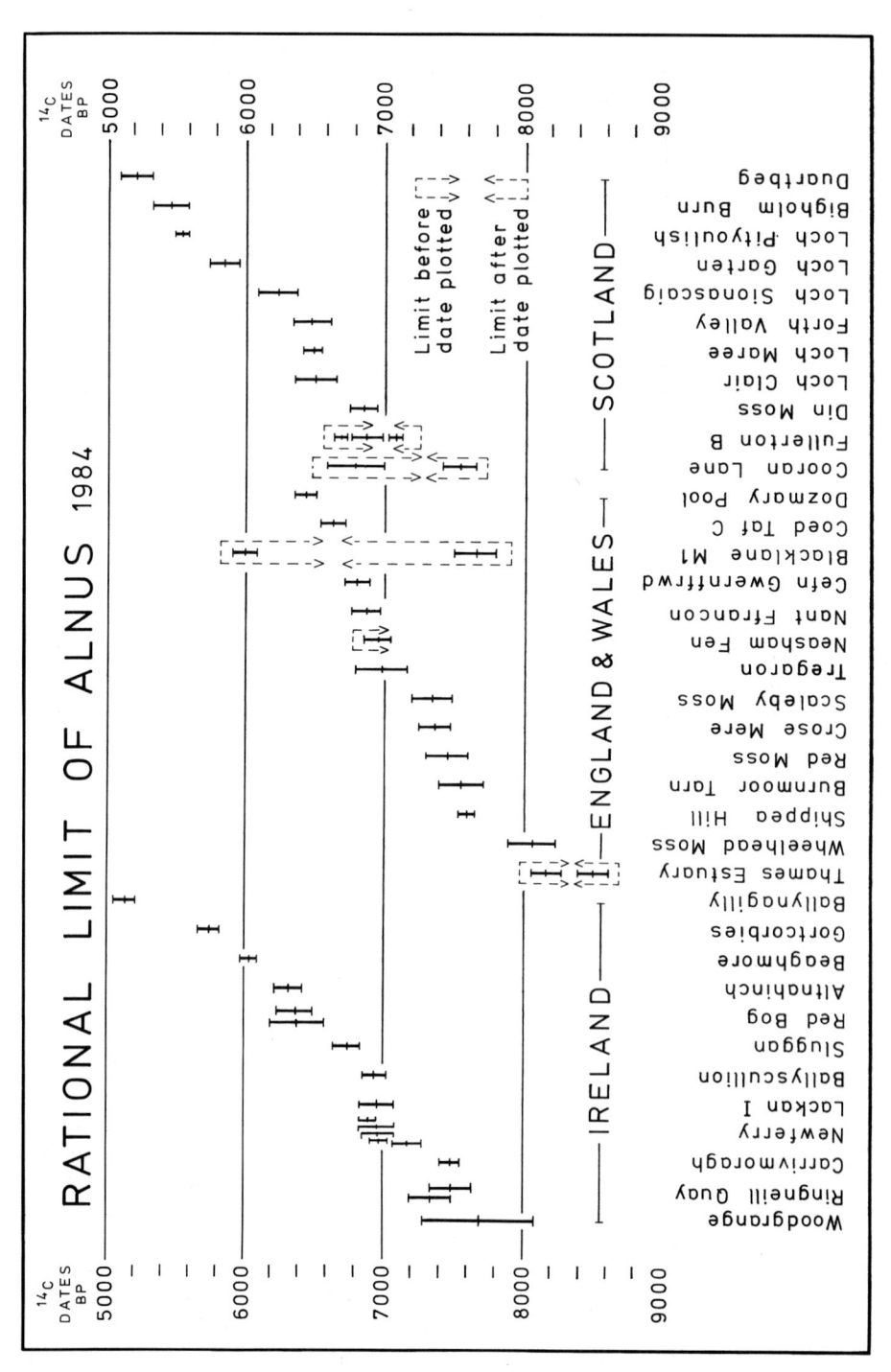

Fig. 1. Conventional radiocarbon dates, with 1σ limits, for the rational limit of *Alnus glutinosa* (L.) Gaertner in the British Isles (see also Appendix).

established in many areas in the Boreal period, before the expansion that often defines the Boreal-Atlantic transition. It would clearly have taken some time for alder to expand into all suitable ecological niches. For Godwin, it is above all, the suddenness of expansion which speaks of climatic effect. And here we come to a different point in the history of alder.

At many sites in Great Britain the expansion of alder does indeed appear to have been very rapid. In Ireland, on the other hand, the expansion is generally slow (Jessen, 1949). These two patterns are not mutually exclusive and I shall give two examples.

(1) The pollen diagram from Wheelhead Moss W.H.2 in Teesdale worked out by Turner *et al.* (1973) shows a very slow rise of the alder curve over a period dated by radiocarbon measurements to some 2800 years (between approximately 8100 and 5800 BP).

(2) In the pollen diagram from Belle Lake in south-east Ireland published by Craig (1978) the alder rise is very sharp. From the ^{14}C-dated deposition rate it took place in approx. 200 years (at approx. 6300 BP).

Where the immigration of alder is early and the rise of the curve slow as in Teesdale, we might suspect a climatic limitation. Conversely, where both the immigration and expansion is late, it might be expected that expansion would be rapid, assuming that – by that time – there would have been no climatic limitation.

Despite the example of Belle Lake, late immigration followed by rapid expansion cannot be held to be true when the pattern in Ireland in general is brought to mind. According to the findings of Huntley & Birks (1983) over much of Ireland (and particularly away from the east coast) the arrival of alder was relatively late, but in general its expansion appears to have been slow.

What then is the reason for the relatively rapid expansion in some areas, and particularly in England? This quesion has been addressed by McVean (1956). In the light of his autecological studies of *Alnus* he gives three possibilities: (1) A rise of sea level producing higher water tables and initiating hydroseres. (2) Increase of precipitation/evaporation ratios leading to moister surface soil in Spring. (3) Destruction of, or interference with, existing vegetation by climatic or biotic factors; e.g. any Mesolithic forest clearance. These effects would have operated by reduction of competition.

Let us examine these three propositions:

(1) Broadly-speaking the major post-glacial sea-level rise was culminating around the Boreal-Atlantic transition (as conventionally dated around 7000/7500 BP). The land bridge between Britain and the continent (perhaps finally between Hull and Esbjerg) was broken around 7800 BP (see Simmons, 1981). Certainly damp coastal habitats would have been created in localities now offshore, but with continually rising sea level they would presumably have been relatively short-lived. Around the maximum of the transgression, however, the worsening drainage of coastal river valleys might have provided habitats long lived enough for alder to have gained a foothold. It appears improbable that any such effect would have been felt at altitudes more than a few metres above sea-level.

(2) In view of the considerations of Huntley & Birks we might, for the moment, exclude the interpretation of the spread of alder itself as indicating wetter conditions at the Boreal-Atlantic transition. We should then be left with somewhat weaker evidence of climatic change. As noted above the evidence of late Boreal dryness is the most compelling. In addition, however, we may recall that evidence from lake geochemistry has been interpreted as pointing to increased wetness

(Pennington *et al.*, 1972). For these reasons we should certainly not abandon the idea of increasing precipitation/evaporation ratios around the transition. We might then take the view that so far as alder is concerned that after some point in time climatic conditions would have been relatively favourable. The end of the late Boreal dryness cannot be dated with any precision. Nevertheless, climatic conditions would presumably have been favourable for alder in western Ireland and north-west Scotland after 6500 to 7000 BP, bearing in mind the generally oceanic climate of these areas. And yet, in general, we see what appears to be a very slow and late process of establishment of alder.

(3) Possible impacts of Mesolithic man at the Boreal-Atlantic transition have been previously discussed by the author and the conclusion reached that vegetational changes that were under way as a result of differential migration rates, soil development and climatic change may have been accelerated by human activity (Smith, 1970, p. 86). It is the purpose of this paper to draw attention to evidence from further investigations and additional sites that strengthens this view and gives additional weight to McVean's third proposition. Emphasis is placed on the evidence from the unique site of Newferry in County Antrim, from which three new pollen diagrams are published.

Newferry

The famous archaeological site at Newferry, Northern Ireland, lies at the northern tip of Lough Beg from which emerges the well known salmon- and eel-supporting river, the Bann. To the south, Lough Beg is connected by a short length of the river to Lough Neagh, the largest body of fresh water in Britain. Lough Neagh receives the drainage of some 5000 to 6000 km^2 of land.

Three archaeological excavations combined with palynological studies have been carried out at Newferry. The most recent, in 1970 to 1971, was carried out by Woodman (1977). Some account of the associated palynological work has been published by Smith (1981). References to earlier work will be found in these two publications.

The major deposit at the site is grey-white diatomite containing both planktonic and benthic species still common in shallow water areas (Battarbee, 1977). From numerous dark, charcoal-containing, occupation layers in the diatomite the water level must have fallen seasonally. Woodman (1977) is of the opinion that this would have been between June and December. The occupation layers in the upper part of the diatomite are Neolithic in age but several Mesolithic occupation layers occur in the lower parts and in the underlying brown reedswamp deposits. The transition between the organic deposits and diatomite is broadly at the Boreal-Atlantic transition. Dating of the occupation layers is not straightforward because of the possibility of redeposition of charcoal. A list of radiocarbon dates is given by Smith, Pearson & Pilcher (1973a, b). These dates, together with some additional determinations not yet published in *Radiocarbon*, are discussed by Woodman (1977) (see also Appendix).

In order to set the scene for the detailed consideration of the Boreal-Atlantic transition that follows, the vegetational history of the site must be rehearsed. The major initial question is the validity of the pollen record from the diatomite: since the water in which the diatoms grew is derived from such a large catchment can the pollen content be considered as representing anything other than this very large area? Contrary to expectation it can be argued that the pollen diagrams represent a relatively small pollen source area. The arguments are presented in detail in Smith & Collins (1971) and Smith (1981). The most compelling reason to believe that this conclusion is correct comes from a comparison of two pollen diagrams one much closer to the valley margin than the other. There are clear parallels but strong differences between the two with much more dramatic changes shown by the site nearer to the dry land. The reason for this apparently local representation is obscure, but presumably connected with the way in which pollen was trapped: possibly largely in very shallow slow-moving water, or even by a tacky diatomite surface that was in the process of drying out.

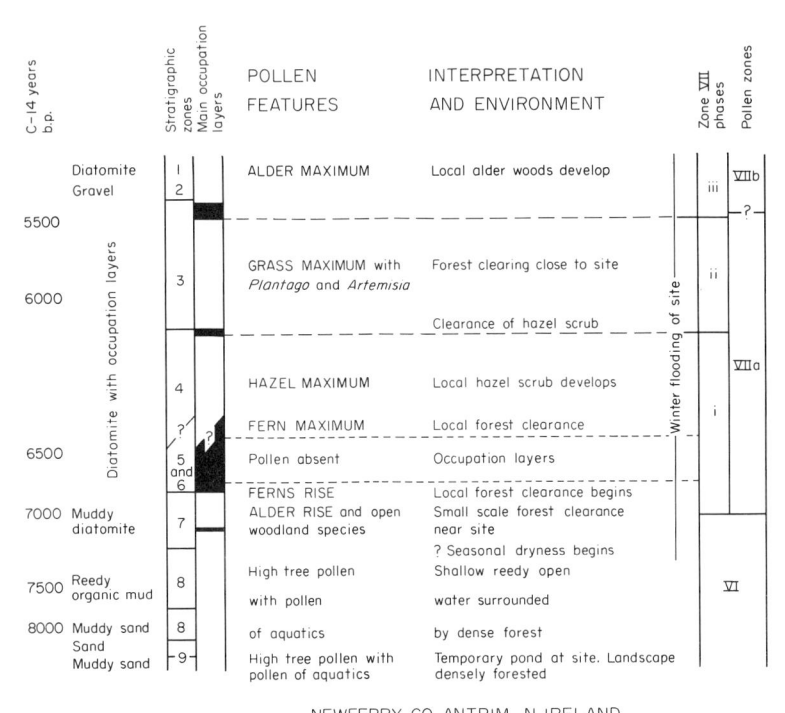

Fig. 3. Schematic diagram (modified after Smith, 1981), based on Figure 2 to show the major features and interpretation of the pollen diagram through the lower diatomite and underlying layers of reedswamp mud and muddy sands at Newferry, Co. Antrim. The timescale, which is approximate only, is based on the ^{14}C dates given in the Appendix. The interpretation of the pollen diagram is mentioned in the text and given in full in Smith (1981).

The diagram referred to as more marginal (from the 1970 to 1971 excavations) is reproduced as Figure 2 and an interpretation (fully argued in Smith, 1981) is summarized in Figure 3. Only a few points need be repeated. Considerable forest clearance appears to have taken place in the Mesolithic period. It was most pronounced around Woodman's stratigraphic zone 6 in the early part of the Atlantic period, pollen zone VIIa. In Figure 3 reference is made to small scale forest clearance in stratigraphic zone 7. The evidence for this will form the major topic of our discussion of the site.

First, however, we must look a little more closely at the dating evidence for the Boreal-Atlantic transition. The dividing line is drawn at the rational limit for *Alnus* within Woodman's stratigraphic zone 7. The ^{14}C dates obtained for this zone are as follows: 6975±120 BP (UB-885); 7190±110 BP (UB-517); 6995±60 BP (UB-516); 6980±115 BP (UB-887); 6915±60 BP (UB-886); 7485±115 BP (UB-497) and 8895±125 BP (UB-636). The first five are consistent and may be taken as dating the stratigraphic zone to approximately 7000 BP [of the last two UB-497 is stratigraphically rather uncertain and UB-636 presumably refers to older wood (see Woodman 1977)]. Alder values reach approx. 10% of the total land-plant pollen in stratigraphic zone 7 and it can hardly be doubted that it became locally well established. The appearance and establishment of alder is earlier than suggested by the isopoll maps of Huntley & Birks. I shall return to this point later.

The detailed pollen diagrams from the lower deposits at Newferry are presented in Figures 4, 5 and 6. It was considered necessary to make several diagrams for two reasons: first, to give some coverage of the stratigraphic variation and secondly, in an attempt to overcome any disturbance of the general sequence by erosion, redeposition or bioturbation.

The first detailed diagram, Monolith II, diagram 2 (Fig. 4) is an enhanced version of Figure 2. The second, Monolith III (Fig. 5) is from a sequence similar to Monolith II, but with a pronounced occupation layer (7, main, of Woodman) at the top of the reedswamp deposit (147 to 150 cm).

The third detailed diagram, Monolith IV (Fig. 6) is from an area of the site where there appeared to have been an attempt to consolidate the surface of the reedswamp deposit (here perhaps more peaty than muddy) by throwing down numerous stones. Note that, (1) The three detailed diagrams have total tree pollen including *Corylus* as the pollen sum whereas in the Monolith II-diagram 1 (Fig. 2) total land-plant pollen is used. (2) The locations of the monoliths in relation to the cuttings of Woodman, are given in the legends to the Figures.

As might be expected there is variation between the diagrams but possibly Monolith II is the least disturbed. It was indeed taken from an area away from the main focus of occupations where charcoal from them appeared to have been gently redistributed by water (see photographs Figs 7, 8). The other two monoliths include layers that would almost certainly have been disturbed during or after deposition. Despite these difficulties a suite of features can be discerned at, or close to the rational limit of the *Alnus* curve which taken together might be interpreted as suggesting some local human influence on the forest cover.

The features are as follows:

(1) A *Quercus* minimum: this can be seen in all three diagrams. At the same level there is a fern-spore maximum (but see below).

(2) Presence or maxima for certain trees or shrubs: in particular, we see maxima in the pollen curves for *Fraxinus* and *Salix* in Monolith II (Fig. 4). In Monolith IV (Fig. 6) there is pollen of *Taxus* and *Populus*. Several of these are present in a more scattered fashion in Monolith III (Fig. 5). Ash is a well known, light-demanding, colonist of woodland clearings. Yew and certain willows can act in a similar way.

(3) Presence of maxima of herbaceous taxa possibly indicative of disturbance or open areas: we may note that (*a*) the consistent presence of *Pteridium* (bracken) spores, with relatively high values in Monolith IV; (*b*) the presence of *Plantago lanceolata* pollen in Monoliths III and IV. *Plantago lanceolata* pollen is, of course, often associated with forest clearances of Neolithic age: note that there are also occurrences below the rational limit of *Alnus* in Monoliths II and III; (*c*) the presence of Compositae pollen, particularly in Monolith IV; (*d*) the presence of *Hyacinthoides* (bluebell) pollen in Monolith IV.

Fig. 7. Photograph, looking west, of cutting G8 of Woodman (1977) from which Monolithic II was taken at Newferry, Co. Antrim. The black occupation layers (redistributed by water) can be seen rising to the South towards the sub-surface sandhill that was the focus of activity in Mesolithic times. Diatomite appears white, or pale grey. The lowest white band (at the level of the top of the iron cladding of the spade shaft) is the upper part of Woodman's stratigraphic zone 7. The top of the spade rests against gravelly deposits of Neolithic age. To the top right of the photograph is a glimpse of the R. Bann.

With reference to (1) it may be noted that the fern spores involved did not belong (morphologically) to marsh species, and that ferns often become extremely abundant nowadays in small woodland clearings. The rise of the fern spore curve appears rather later in Monolith II. As noted above, this is probably the least disturbed and most reliable diagram of the three.

Monolith II, diagram 2, has another feature not strongly exhibited by the other diagrams and that is a temporary decline of *Pinus* (157–154 cm) (Fig. 4). This decline coincides broadly with the *Quercus* decline and *Salix* maximum. It coincides more precisely with the *Fraxinus* maximum. Together with the evidence of disturbance at roughly the same time in the other Monoliths, these features appear to imply some small opening of the forest canopy at the Boreal-Atlantic transition. With the clear evidence of burning of wood on the site by man it is natural to connect such an opening of the canopy with this human activity. It could be argued that there would have been sufficient dead wood available but this is a debatable point, particularly if the occupations lasted for some months each year. On the other hand, if the major activity of prehistoric man at the site was the catching of fish by means of wooden traps and weirs, as suggested by Woodman, then there certainly would have been a requirement for fresh wood in the form of poles and withies. Woodland management for these products is, then, a distinct possibility but must remain as an untestable hypothesis.

The flint and stone tools found at Newferry are considered by Woodman to be the maintenance equipment for fish weirs and traps. That is to say, they are essentially woodworking tools. It is of interest to note that the tool kit included polished stone axes even in Mesolithic times. The level of the oldest of the polished stone axes (in stratigraphic zone 8) is indicated in Figure 5, the monolith having been taken precisely at the find spot. The level of the axe coincides within 1 cm

Fig. 8. Detail of the section in Figure 7 with a 50 cm monolith box placed in the position from which Monolith II was secured (Figs 2 and 4). The thin black layer in the middle of the white diatomite is at the top of Woodman's (1977) stratigraphic zone 4 (see 120 cm in Fig. 2). The thick band of laminated black and white layers (whose base is at the junction between the full white and black sections of the vertical ranging pole) are Woodman's stratigraphic zones 5 and 6. The white layer immediately below these is the upper part of Woodman's stratigraphic zone 7 (approx. 150–158 cm. in Fig. 2).

with the first occurrence of *Plantago lanceolata* pollen. If the axe happened to have been pushed down in the deposit this correspondence could be coincidental but the early presence of polished stone axes does add some realism to the suggestion made above.

The rational limit of the *Alnus* curve in the Newferry Monolith II diagram (Fig. 4) occurs just below the *Pinus* decline and *Fraxinus* maximum. The immigration of alder does not then appear to be necessarily connected with the disturbance of woodland cover but the spatial significance of low *Alnus* values is imponderable. If the arguments above are accepted, however, the broad coincidence of the beginning of the alder curve with signs of disturbance of the existing forest cover may well be of significance. The establishment of alder in a disturbed riverain habitat would accord well with the observation that can readily be made today of the establishment of alders at flood height along stream banks where their seed must have been stranded. The reduction of competition by human disturbance of the existing vegetation cover could then have been a key factor.

At the top of the Monolith II diagram 2 (Fig. 4) we see increasing values of alder and the decline of the curves of other arboreal taxa save for hazel. At the same time there begins the massive rise of the fern-spore curve which has been interpreted (above and Smith, 1981) as probably indicating the creation of further open areas that were eventually colonized by hazel (see Figs 2 and 3). A secondary hazel maximum at, or close to, the Boreal-Atlantic transition is not uncommon in British pollen diagrams. Several of the examples are cited by Smith (1970) where it is pointed out that one of the possible explanations of this feature is an opening of the forest canopy by man. Without repeating in detail the arguments previously published (Smith, 1981) it is worth passing note here that the very strong rise of the alder curve between phases (ii) and (iii) in the Monolith II, diagram 1 (Fig. 2) may also be a reaction to human influence. It comes at the end of a well defined phase (ii) with very high grass pollen values with disturbance indicators such as *Plantago lanceolata* and *Artemisia* which has been interpreted as possibly reflecting the development of open areas (conceivably maintained by grazing) following removal of the hazel scrub of late Phase (i).

NEWFERRY IN THE IRISH CONTEXT

The significance of the general depositional change at Newferry from reedswamp deposits to diatomite has been discussed by Smith (1970, 1981). It is pointed out that a wide scale removal of forest cover in the catchment in Mesolithic times could have affected the hydrology of the Lower Bann Valley but that, bearing in mind the large scale of the event, it seems unlikely that prehistoric man was a major factor. On the other hand, the change in the hydrological regime is unlikely to have been connected with changes of sea level. Even though the post-glacial sea level rise was culminating at about this time (Carter, 1982), the steep fall of the river at The Cutts, Coleraine, would have prevented any change in the drainage base of the major portion of the River Bann. It could be argued that the change to diatomite deposition at Newferry – which took place in Late Boreal times – is indicative of generally wetter conditions. That is, that the depositional change supports the classical view of climatic change. If that were the case we might expect to see similar changes elsewhere. Aside from the Lower Bann valley, however, diatomite is an infrequent type of deposit. There are records from Ballynagilly, Co. Tyrone (Pilcher & Smith 1979) and Jenkinstown, Co. Louth (Mitchell, 1951). The former occurrence is early post-glacial and the latter probably even earlier. At Belle Lake, Co. Waterford (Craig, 1978) there is a more relevant diatomite deposite. This occurs as a thin layer within organic muds and comes immediately below the first sharp rise of the *Alnus* curve. There is little in the pollen diagram to suggest any connection with forest clearance but that possibility certainly exists.

There appears, then, to be little general evidence for flooding at the Boreal-Atlantic transition. Indeed, stratigraphic evidence from lowland raised bogs in north-east Ireland may be found to oppose the idea of increased climatic wetness. In some cases the transition is the point at which fen peats give way to acid bog peats. The top of the reedswamp deposit, was, however, invaded by pines which appear to have persisted during the initial stages of acid peat accumulation (e.g. Sluggan Moss, Co. Antrin (I. C. Goddard, 1971) and Fallahogy, Co. Londonderry (Smith, 1958 and subsequent unpublished observations). The invasion of pine at these sites may be simply a matter of autogenic succession. On the other hand, it might indicate dryness either locally or more generally. Clearly, a more detailed study of tree layers of Boreal-Atlantic age Ireland is required, but we may note that Birks (1972a) has already concluded that in the Galloway region of Scotland conditions in her *Alnus-Ulmus* zone (corresponding in part to the Atlantic period) may have been less oceanic than today.

Estuarine Habitats at the Boreal-Atlantic Transition

We have seen that the Newferry site is not likely to have been affected by sea-level changes. In the estuary of the River Bann, however, the alder curve begins a short time after the first indications of marine influence (Bann Estuary pollen diagram; Jessen, 1949). It begins during a stage of brackish water sediment deposition. The brackish water sediment was laid down over a beach sand which itself formed above a wood peat. These events are unfortunately not closely dated. At Woodgrange, in Co. Down, however, we see the beginning of the alder curve at a level dated to 7650 ± 400 BP (Singh & Smith, 1973) in a coastal basin first affected by marine inundation somewhat earlier. It was previously thought (Singh & Smith, 1973, p. 34) that there was no reason to suppose that the marine inundation at this site brought about the early spread of alder (as suggested by Mitchell, 1951) because of the slow rise of the alder curve. Now that we see the ^{14}C date for the appearance of alder at Woodgrange in a broader context (Fig. 1) and that, despite the large probable error, it is an early date, we can perhaps revert to the idea of coastal areas such as this as being the habitats in which alder first gained a foothold. We shall see (below) that there is other evidence of the early establishment of alder in Co. Down. It may be added here, however, that the dates of approx. 7350 to 7500 BP (Q–632) for another maritime habitat in Co. Down at Ringneill Quay also suggest the early establishment of alder. These dates were obtained for a lagoon clay in which the *Alnus* curve becomes continuous and rises to 3 to 5% of the total tree pollen (Morrison, 1961; see also Appendix).

While the evidence is perhaps not yet complete enough to draw an unassailable general conclusion, it is of interest to note two further pieces of evidence regarding the importance of estuaries in the establishment and spread of alder.

Firstly, alder was apparently well established in the Thames Estuary by 8170 ± 110 BP (Q–1426) (Devoy, 1979). Evidence from the Thames gives the bridgehead for immigration into south-east England in the maps of Huntley & Birks.

Secondly, in eastern Scotland, north of the Tay, at Fullerton near Montrose, D. E. Smith *et al.* (1980) have shown that the rational limit of the alder curve comes within the estuarine carse clay deposit. The clay deposit is bracketed by ^{14}C dates of 6704 ± 110 BP (SRR–1148), above, and by two dates 7086 ± 50 BP (SRR–1149) and 6880 ± 110 BP (Birm–867) below. Alder appears thus to have become established there around 7000 BP. However the ^{14}C determinations are interpreted, this is the earliest establishment of alder in Scotland aside from the more southerly (and western) area of Galloway (see Fig. 1).

The evidence thus begins to suggest that in the three regions estuaries were the places of earliest establishment of alder. Further dates from similar locations will obviously be of much interest. Meanwhile, we may note that, in the Forth Valley, west of Stirling, a record of the Boreal-Atlantic transition at 6490 ± 125 BP occurs at an interface of peat above the carse clay, the deposition of which began in the same area about a thousand years earlier (Sissons & Brooks, 1971). The expansion of alder there may be connected with the creation of suitable new habitats, initially free from competition, as the sea receded.

Human Influence at the Boreal-Atlantic Transition

Some general discussion of possible human effects at the Boreal-Atlantic transition has been given by Smith (1970). There are a number of sites, however, which have particular parallels with Newferry and a summary of these will allow us to broaden our conclusions.

Sites in Co. Down

The idea of an early arrival of alder in north-east Ireland as deduced for Newferry (and sites in Co. Down) receives support from the findings of the late Dr S. M. Holland in County Down. In her diagram from Lackan I (Holland, 1975) the rational border for *Alnus* is dated to 6975±110 BP (UB-803, see Fig. 1), though the main expansion is later. An even earlier date of 7495±70 BP (UB-872) was obtained for the rational border of *Alnus* for her site at Carrivmoragh [see Fig. 1, for details see Pearson & Pilcher (1975)]. The latter date is, of course, very close to that from Woodgrange noted above (7650±400 BP). It is possible then that the immigration of alder into Ireland began in the north-east.

It is of much interest to note at the Lackan site that, at the rational border of the *Alnus* curve, there is a sample with very high alder and low *Pinus* values coming some time before the general rise of the curve. Immediately below (and also at approx. 6300 BP alongside the later general rise) Holland encountered single pollen grains of *Plantago lanceolata*. As at Newferry, *Taxus* pollen is also recorded. Holland recorded the presence of charcoal at the same levels, though charcoal appears from her stratigraphic representations to also have had a wider occurrence. Holland did not discount the possibility that the plantain pollen might be linked with Mesolithic man who is known to have been present in Co. Down at the time. In view of the consideration of the Newferry site it now seems that the connection can be made more strongly. Indeed, the temporary *Alnus* peak may be more than coincidental though it does not show up in a detailed analysis of the horizon (Holland's Lackan Horizon I diagram).

Sites in England and Wales

The particularly striking feature of the Newferry sequence in the field and clearly visible in the photographs (Figs 7 and 8) is the black charcoal-containing layers between the basal reedswamp deposits and the overlying diatomite.

Distinct black layers of similar biostratigraphic age are known from two other sites. The first is the classical East Anglian site of Shippea Hill (Clark, Godwin & Clifford, 1935; Godwin & Clifford, 1938; Clark & Godwin, 1962). The thick black Mesolithic layer there, dated to 7600±150 BP (Q-587) falls exactly at the Boreal-Atlantic transition. Unfortunately, pollen preservation was poor in this layer and the diagram lacks detail. It is of interest, however, to note that it contains a fern-spore maximum, as at Newferry. A more detailed new investigation of this layer elsewhere on the site would be of much interest.

New investigations in the Vale of Pickering, East Yorkshire, close to the famous Mesolithic site of Star Carr have revealed yet another black layer associated with the Boreal-Atlantic transition (Cloutman & Smith, unpublished). The layer begins at a point where alder pollen has been present for a few centimetres and its lower limit coincides with a sharp fall of the pine curve and rise of alder. Starting at the same level as the beginning of the *Alnus* curve are certain features which suggest opening of forest cover, and soil disturbance. These are increased grass pollen values and the occurrence of *Artemisia* and Chenopodiaceae pollen. The date of the transition at this site has not yet been determined. Indeed, so far as can be ascertained, no radiocarbon dates have been published for the Boreal-Atlantic transition in East Yorkshire.

We may also recall the possible evidence of disturbance of the forest at the beginning of the Atlantic period again with some evidence of burning at two Yorkshire sites in the Pennines: Malham Tarn (Pigott & Pigott, 1963) and Stump Cross (Walker, 1956) [for a more complete discussion see Smith (1970)]. At Stump Cross carbonized wood associated with Mesolithic flint artefacts yielded a radiocarbon date of 6500 ± 310 BP (Q-141) (Godwin & Willis, 1962; this date is not 8450 ± 300 BP as stated by Switsur & Jacobi, 1975).

The Teesdale site of Dead Crook Moss (DC I) (Turner et al., 1973) provides an extremely interesting parallel to Holland's Co. Down site of Lackan mentioned above. At the end of the authors' zone H, probably datable to around 8000 BP, there is a single sample above the empirical limit of the alder curve in which the alder value reaches almost 20% of the total tree pollen. This is tentatively interpreted as possibly implying a brief period of local alder growth (Turner et al. 1973, p. 399). The interest of this high alder value for the present discussion is that the same sample contains *Plantago lanceolata* pollen. This is not a single grain representation as at Lackan, but at a level of some 2 to 3% of the total tree pollen (so far as this can be read from the published diagram). This coincidence might be taken as confirming the emerging suspicion that alder could have reacted – more generally than at Newferry – as a colonist of cleared areas in the Mesolithic. For the moment, however, any such implication must be treated with caution since a similar high alder value with associated *Plantago lanceolata* was found by Turner et al. at Red Sike Moss (RS II) in a Late-Devensian context! Detailed analysis of *Alnus* peaks coming before the main expansion would, however, obviously be of much interest in future work.

At Dozmary Pool (Bodmin Moor, Cornwall) the alder curve rises at a level dated to 6451 ± 65 BP (Q–1025) at the junction between zones DP 5 and DP 5a of Brown (1977) where the sedge-peat gives way to raised bog peat. In zone DP 5, Brown argues that woodland on better drained soils was restricted by fire. In zone DP 5a, he sees the establishment of alder as a carr tree and this as being connected with the onset of a wetter climate. The change from sedge peat to raised bog peat Brown attributes to the same cause. At both this site and his Hawks Tor site we may note, however, the beginning of a continuous *Plantago lanceolata* curve as the alder curve begins.

More extensive signs of interference with woodland cover, again associated with charcoal have been found on Dartmoor at Blacklane (Simmons, 1964; Simmons, Rand & Crabtree, 1983). The authors demonstrate in their zone BLB 5 (presumably mis-numbered as 6 in their fig. 4) a suite of taxa usually associated with the opening of forest. These taxa include *Pteridium* and other polypodiaceous fern spores. At the same level the frequency of charcoal and of *Fraxinus* pollen increases. The phase is bracketed by ^{14}C dates of 7760 ± 140 BP (HAR-4462) and 6010 ± 90 BP (Har-4461) (see Fig. 1). It is interpreted as one of minor forest recession with grassland replacing some deciduous woodland most likely as a result of human activity. The authors suggest that fire may have been used by Mesolithic people to maintain open ground in landscape that was predominantly mature forest.

What is of most interest for the present discussion is that the alder curve begins to rise from its rational border during the later part of this disturbance. The fern spore maximum and the presence of *Fraxinus* are quite close parallels with the sequence at Newferry. An interpolated date for the rational limit of alder at Blacklane is approx. 6800 BP. This is of course quite close to the date of approx. 7000 BP for the beginning of the alder rise at Newferry.

Pennington et al. (1972) interpret increases of iron and manganese in lake sediments in north-west Scotland, in particular at Loch Sionascaig, and changes in the diatom flora at the Boreal-Atlantic transition (there dated to approx 6250 BP) as possibly due to an increase in rainfall and rising water tables leading to increased erosion of soils. Their primary conclusion is, however, of increased solutional transport from waterlogged (reducing) soils (Pennington, 1981).

I have already alluded to the possibility of soil erosion induced by man at the Boreal-Atlantic transition. This discussion will be brought to a close by descriptions of two sites showing effects of Mesolithic man on soils. At the first site there is evidence – albeit on a quite local scale – of physical disturbance. The site, at the Breidden in mid-Wales has archaeological importance. It was excavated by C. Musson & W. A. Britnell who discovered within a Hillfort a small pond deposit dating back to the Devensian Late-glacial. The pond deposits, which have been investigated pollen analytically (Smith & Green, unpublished) contained large upturned turves of soil (identified as such by S. Limbrey). The pollen analytical investigations show that at least one such soil turf – which can hardly have found its way into the pond other than at the hand of man – almost certainly belongs to the Boreal-Atlantic transition, there dated to 6895 ± 85 BP (CAR-157), or even a little earlier. Such minor soil disturbances can hardly have altered the run-off from whole catchments but at the second site, also in Wales, considerable changes in soil conditions appear to have been brought about by the activities of man in the Mesolithic period.

This site in question is at Waun Fignen Felen in the Black Mountain range of South Wales (Smith & Cloutman, unpublished). Here is a small Late-Devensian lake basin which was eventually overgrown by the blanket peats which mantle the surrounding drift. At the base of the peat, on the drift, several Mesolithic working floors have been uncovered. The basal blanket peat is commonly in the form of a greasy, black (charcoal containing) mor. Fourteen sites have been investigated by means of pollen analysis and ^{14}C dating. The oldest blanket peats began to form at approx. 7600 BP but more generally in the few hundred years around 6100 BP. In one area of the blanket bog the basal mineral soil is podsolized. Here, it is apparent that the black greasy mor deposit went on accumulating for about 1000 years until true blanket peat started to accumulate at approx. 5500 BP. From the pollen content, the mor accumulated under heath which, considering the high charcoal content, appears to have been maintained by repeated burning. The presence of Mesolithic artefacts provides strong circumstantial evidence that this burning was at the hand of man.

Podsolization would hardly have encouraged the spread of alder, but it is of interest to note that the rational limit of *Alnus* at Waun Fignen Felen occurs once more at a time of apparent disturbance, in the later-dated of the black mor layers around 6100 BP, though earlier in the lake basin.

The podsol at Waun Fignen Felen is one of the oldest fossil podsols known from Britain. Bearing in mind the effects of Mesolithic man on soils which have been detected by Dimbleby (e.g. Rankine, Rankine & Dimbleby, 1960) we may conclude that it probably formed largely as a result of human activity. If on a wide enough scale, the process of podsolization would presumably be reflected in the geochemistry of sediments in lakes receiving the drainage of the catchment. Geochemical work on lake sediments in Wales is, however, in its infancy.

We now have at least four sites at which we see the rise of the alder curve associated with signs of burning and woodland clearance. These are Newferry, Blacklane, Seamer Carr and Waun Fignen Felen. The geographical and altitudinal range of these sites does much to make more general the idea of human effect in the spread of alder in Britain. This is not to deny that alder might have become established by natural means and indeed that in many areas it probably did so. Nevertheless, it now seems clear that man could have influenced the time at which alder became established or at which it expanded. The extent to which he may have affected the rate of expansion; that is, the rapid English – as compared with the slow Irish – rise of the curve, must remain uncertain.

Conclusions

Reverting to the three possibilities relating to the spread of alder raised by McVean (1956), our conclusions must be as follows.

Rise of sea level

Alder [*Alnus glutinosa* (L.) Gaertner] does appear to have become established relatively early in coastal areas affected by marine influences. Estuaries may have been primary sites of invasion. The establishment of alder may have been facilitated by the damaging effects of the marine influence on the existing valley floor plant communities, or by the creation of new habitats.

Increase of precipitation/evaporation ratios

Where the establishment of alder was relatively early it may well have benefited from the contemporaneous ending of a phase of climatic dryness. The expansion of alder, particularly where relatively late, should not itself be taken necessarily as indicative of a climatic change. The late arrival and expansion of alder in northern and western Britain may be seen primarily as a natural process of migration.

Destruction of, or interference with, existing vegetation by climatic or biotic factors

Early climatic influence, particularly the Boreal dry phase probably did have effects in disturbing vegetation cover (see particularly Brown, 1977). The ending of this phase can less easily be seen as destructive. The influence of Mesolithic man can, however, be seen in exactly this way. By reduction of local competition the destructive effects appear to have assisted the establishment of alder. If the secondary hazel maximum at the Boreal-Atlantic transition, which is even more widespread in Britain than the evidence of burning etc. noted in this discussion, is accepted as due to woodland destruction by Mesolithic man (see Smith, 1970), then human influence in the expansion of alder may have been considerable.

Acknowledgements

The pollen diagrams in Figures 5 and 6 were prepared by Mr P. Medhurst in Belfast, and those in Figures 2 and 4 by Mrs E. A. Brown in Cardiff. The author, who checked the critical identifications, expresses his gratitude for their painstaking work. Thanks are also due to Miss L. A. Morgan for assistance in preparing this paper, in particular for drawing Figure 1. Other acknowledgements of help and collaboration during the Newferry excavations are given in Smith (1981).

Appendix

Annotated list of radiocarbon dates for the rational limit of the *Alnus* curve (quoted as BP) plotted in Figure 1 and for the archaeological layers at Newferry. In selecting the dates an attempt has been made to define the point at which alder appears to have begun to become substantially established. The exact point at which this occurs is sometimes a matter of subjective assessment. The radiocarbon samples are often relatively thick, and occasionally not precisely related to the horizon defined. For these reasons the use of Figure 1 should not supplant examination of the original sources. Original publication of the radiocarbon dates may be traced from the indices to Laboratory code numbers (which are given where possible) in *Radiocarbon*. References in the list are mainly to the publication containing the pollen diagram.

Ireland

Woodgrange, Co. Down. Singh & Smith (1973). 7650±400 (LJ-904).

Ringneill Quay, Co. Down. Morrison (1961). 7345±150, 7500±160 (Q-632).

Note: the radiocarbon sample was 8 to 10 cm below the top of a lagoon clay (Godwin & Willis, 1962). This level corresponds closely with the slight rise of the *Alnus* curve to consistent values of 4 to 5% of the total tree pollen shown in Morrison's pollen diagrams.

Carrivmoragh, Co. Down. Holland (1975). 7495±70 (UB-872).

Newferry, Co. Antrim. Radiocarbon dates from the archaeological layers (stratigraphical zones). The numbers of the date-groups are those given in Figure 2. Except where stated the dates were obtained from charcoal.

(1) Upper zone 3: 5310±85 (UB-503), 5415±95 (UB-489), 5705±90 (UB-630), 5795±105 (UB-508).

(2) Zone 4 (occupation layer at junction with zone 3): 6215±100 (UB-490).

(3) Zone 5: 6640±185 (UB-653), 6605±175 (UB-505).

(4) Zone 6: 7175±105 (UB-514); (diatomite with organic mud and charcoal): 6590±60 (UB-380).

(5) Zone 7 (upper). 7075±125 (UB-885); main: 6915±60 (UB-886), 6955±160 (UB-516); 6930±115 (UB-887); 7075±125 (UB-885); 7190±110 (UB-517) plus two others of uncertain relevance (see text): 7485±115 (UB-496); 8995±125 (UB-637).

(6) Zone 8 (wood): 7360±195 (UB-641).

(7) Zone 9 (organic mud): 8175±70 (UB-888); (wood): 8190±120 (UB-487).

(Where there is variation of detail from dates or laboratory codes previously published (Smith *et al.*, 1973a, b; Woodman, 1977) this list – which has been checked with Dr G. W. Pearson – is to be regarded as definitive).

Lackan I, Co. Down. Holland (1975). 6975±110 (UB-803).

Ballyscullion, Co. Antrim. Smith *et al.* (1971). 6950±85 (UB-120). Note: sample very slightly above rational limit.

Sluggan, Co. Antrim. Smith (1979); Smith *et al.* (1971). 6760±90 (UB-120).

Red Bog, Co. Louth. Watts, unpublished: Mitchell (1956); McAulay & Watts (1961). 5170±190 (D-4).

Altnahinch, Co. Antrim. A. Goddard (1971). 6340±100 (UB-422).

Beaghmore, Co. Tyrone. Pilcher (1969). 6050±60 (UB-94).

Gortcorbies, Co. Tyrone. I. C. Goddard (1971). 5160±75 (UB-234).

Ballynagilly, Co. Tyrone. Pilcher & Smith (1979). 5145±70 (UB-253).

England and Wales
Thames Estuary. Devoy (1979). 8510±110 (Q-1286) from the Isle of Grain, EB 135, is below, and 8170±110 (Q-1426) from Tilbury is above the rational limit.

Wheelhead Moss, Upper Teesdale. Turner et al. (1973) 8070±170 (GaK-2917).

Shippea Hill, Cambridgeshire. Clark & Godwin (1962). 7610±150 (Q-587).

Burnmor Tarn, Cumbria. Pennington (1970). 7560±160 (Y-2361).

Red Moss, Lancs. Hibbert, Switsur & West (1971). 7460±150 (Q-917).

Crose Mere, Shropshire. Beales (1980). 7373±110 (Q-1236). Note: date is somewhat above the rational limit.

Scaleby Moss, Cumbria. Godwin, Walker & Willis (1957). 7361±146 (Q-167). Note: sample is above the rational limit.

Tregaron, Dyfed. Hibbert & Switsur (1976). 6990±130 (Q-937). Note. *Alnus* pollen appears earlier, below a level dated to 7130±180 (Q-936).

Neasham Fen, Durham. Bartley, Chambers & Hart-Jones (1976). 6962±90 (SRR-103). Note: date is slightly above the rational limit.

Nant Ffrancon, Gwynedd. Hibbert & Switsur (1976). 6880±110 (Q-900). Note: *Alnus* pollen is present much earlier, down to a level dated to 8450±150 (Q-898).

Cefn Gwernffrwd, Powys. Chambers (1982). 6815±85 (CAR-96).

Blacklane M1, Cornwall. Simmons et al. (1983). 6010±90 (HAR-4461) and 7660±140 (HAR-6642) bracket the rational limit which by interpolation appears to be at approx. 6800 BP.

Coed Taf C, Mid-Glamorgan. Chambers (1983). 6645±85 (CAR-89).

Dozmary Pool, Cornwall. Brown (1977). 6451±65 (Q-1025).

Scotland
Cooran Lane, Galloway. Birks (1975). 6805±1200 (Q-873) above, and 7541±120 (Q-874) below rational limit.

Fullerton, former County of Angus. D. E. Smith et al. (1980). 6704±55 (SRR-1148) above; 7086±50 (SRR–1149) and 6880±110 (Birm-867) below rational limit.

Din Moss, former County of Roxburghshire. Hibbert & Switsur (1976). 6860±100 (Q-1069).

Loch Clair, former County of Wester Ross. Pennington et al. (1972). 6520±145 (I-4814).

Loch Maree, Former County of Wester Ross. Birks (1972b). 6513±65 (Q-1007).

Fourth Valley. Sissons & Brooks (1971). 6490±125. Note: sample is described as referring to the Boreal-Atlantic transition but could be above the rational limit of alder.

Loch Sionascaig, former County of Wester Ross. Pennington et al. (1972). 6250±140 (Y-2363).

Loch Garten, Speyside. O'Sullivan (1974). 5860±100 (UB-851).

Loch Pityoulish, Speyside. O'Sullivan (1976). 5548±50 (SRR-459).

Bigholm Burn, former County of Dumfriesshire. Moar (1969a). 5475±120 (Q-702).

Duartbeg, former County of Sutherland. Moar (1969b). 5220±115 (Q-748).

REFERENCES

ANDERSSON, G. (1909). The climate of Sweden in the Late-Quaternary period. *Sveriges Geologiska Undersøkning*, Ser.C., Arsbok 3, **1**, 1–88.
BARTLEY, D. D., CHAMBERS, C. & HART-JONES, B. (1976). The vegetational history of parts of south and east Durham. *The New Phytologist*, **77**, 437–468.
BATTARBEE, R. (1977). *Diatom Analyses of a Newferry Monolith*. Appendix I in Woodman, P. C. (1977) (see below).
BEALES, P. W. (1980). The Late-Devensian and Flandrian vegetational history of Crose Mere, Shropshire. *The New Phytologist*, **85**, 133–161.
BIRKS, H. H. (1972a). Studies in the vegetational history of Scotland. II. Two pollen diagrams from the Galloway Hills, Kirkudbrightshire. *Journal of Ecology*, **60**, 183–217.
BIRKS, H. H. (1972b). Studies in the vegetational history of Scotland. III. A radiocarbon dated pollen diagram from Loch Maree, Ross and Cromarty. *The New Phytologist*, **71**, 731–754.
BIRKS, H. H. (1975). Studies in the vegetational history of Scotland. IV. Pine stumps in Scottish blanket peats. *Philosophical Transactions of the Royal Society of London* B, **270**, 181–226.
BROWN, A. P. (1977). Late-Devensian and Flandrian history of Bodmin Moor, Cornwall. *Philosophical Transactions of the Royal Society of London* B, **276**, 251–320.
CARTER, R. W. G. (1982). Sea-level changes in Northern Ireland. *Proceedings of the Geologists' Association*, **93**, 7–23.
CHAMBERS, F. M. (1981). Date of blanket peat initiation in upland South Wales. *Quaternary Newsletter*, **35**, 24–29.
CHAMBERS, F. M. (1982). Environmental history of Cefn Gwernffrwd, near Rhandirmwyn, Mid-Wales. *The New Phytologist*, **92**, 607–615.
CHAMBERS, F. M. (1983). Three radiocarbon-dated pollen diagrams from upland peats north-west of Merthyr Tydfil, South Wales. *Journal of Ecology*, **71**, 475–487.
CLARK, J. G. D. & GODWIN, H. (1956). A Maglemosian site at Brandesburton, Holderness, Yorkshire. *Proceedings of the Prehistoric Society*, **22**, 6–22.
CLARK, J. G. D. & GODWIN, H. (1962). The Neolithic in the Cambridgeshire Fens. *Antiquity*, **36**, 10–23.
CLARK, J. G. D., GODWIN, H. & CLIFFORD, M. H. (1935). Report on recent excavations at Peacock's Farm, Shippea Hill, Cambridgeshire. *Antiquaries Journal*, **15**, 284–319.
CRAIG, A. J. (1978). Pollen percentage and influx analysis in south-east Ireland: a contribution to the ecological history of the Late-glacial period. *Journal of Ecology*, **66**, 297–324.
DEVOY, R. J. N. (1979). Flandrian sea level changes and vegetational history of the Lower Thames Estuary. *Philosophical Transactions of the Royal Society of London* B, **285**, 355–407.
FRIES, M. (1965a). Outlines of the Late-glacial and postglacial vegetation and climatic history of Sweden, illustrated by three generalized pollen diagrams. *International Studies on the Quaternary. Papers of the VII Congress of the International Association for Quaternary Research, Boulder, Colorado, 1965*. Geological Society of America. Special Paper 84 (Ed. by H. E. Wright, Jr & D. G. Frey), pp. 55–64.
FRIES, M. (1965b). The Late-Quaternary vegetation of Sweden. *Acta Phytogeographica Suecica*, **50**, 269–284.
GODDARD, A. (1971). *Studies of the vegetational changes associated with the initiation of blanket-peat accumulation in north-east Ireland*. Unpublished Ph.D. thesis, Queen's University, Belfast.
GODDARD, I. C. (1971). *The palaeoecology of some sites in the north of Ireland*. Unpublished MSc. thesis, Queen's University, Belfast.
GODWIN, H. (1940a). Pollen analysis and forest history of England and Wales. *The New Phytologist*, **39**, 370–400.
GODWIN, H. (1940b), Studies on the post-glacial history of British vegetation. III. Fenland pollen diagrams. IV. Post-glacial changes of relative land and sea-level in the English Fenland. *Philosophical Transactions of the Royal Society of London* B, **230**, 239–303.
GODWIN, H. (1975). *The History of the British Flora*. 2nd edn. Cambridge University Press, Cambridge.
GODWIN, H. & CLIFFORD, M. H. (1938). Studies of the post-glacial history of British vegetation. I. Origin and stratigraphy of Fenland deposits near Woodwalton, Hunts. II. Origin and stratigraphy of deposits in southern Fenland. *Philosophical Transactions of the Royal Society of London* B, **229**, 323–406.
GODWIN, H. & TALLANTIRE, P. A. (1951). Studies in the post-glacial history of British vegetation. XII. Hockam Mere, Norfolk. *Journal of Ecology*, **39**, 285–307.
GODWIN, H., WALKER, & WILLIS, E. H. (1957). Radiocarbon dating and post-glacial history: Scaleby Moss. *Proceedings of the Royal Society of London* B, **147**, 352–366.
GODWIN, H. & SWITSUR, V. R. (1959). Cambridge University natural radiocarbon measurements I. *American Journal of Science Radiocarbon Supplement*, **1**, 63–75.
GODWIN, H. & WILLIS, E. H. (1962). Cambridge University natural radiocarbon measurements V. *Radiocarbon*, **4**, 57–70.
HIBBERT, F. A. & SWITSUR, V. R. (1976). Radiocarbon dating of Flandrian pollen zones in Wales and northern England. *The New Phytologist*, **77**, 793–807.

HIBBERT, F. A., SWITSUR, V. R. & WEST, R. G. (1971). Radiocarbon dating of Flandrian pollen zones at Red Moss, Lancashire. *Proceedings of the Royal Society of London* B, **177**, 161–176.

HOLLAND, S. M. (1975). *A pollen analytical study concerning settlement and early ecology in Co. Down, Northern Ireland*. Unpublished Ph.D thesis: Queen's University, Belfast.

HUNTLEY, B. & BIRKS, H. H. (1983). *An Atlas of Past and Present Pollen Maps for Europe: 0–13000 years ago*. Cambridge University Press, Cambridge.

JESSEN, K. (1949). Studies in Late-Quaternary deposits and flora-history of Ireland. *Proceedings of the Royal Irish Academy*, **52B**, 85–290.

MCCAULAY, I. R. & WATTS, W. A. (1961). Dublin radiocarbon dates I. *Radiocarbon*, **3**, 26–38.

MCVEAN, D. N. (1956). Ecology of *Alnus glutinosa* (L.) Gaertn. VI. Post-glacial history. *Journal of Ecology*, **44**, 331–333.

MITCHELL, G. F. (1951). Studies in Irish Quaternary deposits: No. 7. *Proceedings of the Royal Irish Academy*, **53B**, 111–206.

MITCHELL, G. F. (1956). Post-Boreal pollen diagrams from Irish raised bogs. (Studies in Irish Quaternary deposits: No. 11). *Proceedings of the Royal Irish Academy*, **57B**, 185–251.

MOAR, N. T. (1969a). A radiocarbon-dated pollen diagram from north-west Scotland. *The New Phytologist*, **68**, 209–214.

MOAR, T. N. (1969a). Late Weichselian and Flandrian pollen diagrams from south-west Scotland. *The New Phytologist*, **68**, 433–467.

MOORE, P. D. (1975). Origins of blanket mires. *Nature*, **256**, 267–269.

MORRISON, M. E. S. (1961). The palynology of Ringneill Quay, new Mesolithic site in Co. Down, Northern Ireland. *Proceedings of the Royal Irish Academy*, **61C**, 171–182.

O'SULLIVAN, P. E. (1974). Two Flandrian pollen diagrams from the East Central Highlands of Scotland. *Pollen et Spores*, **16**, 33–57.

O'SULLIVAN, P. E. (1976). Pollen analyses and radiocarbon dating of a core from Loch Pityoulish, eastern Highlands of Scotland. *Journal of Biogeography*, **3**, 293–302.

PEARSON, G. W. & PILCHER, J. R. (1975). Belfast Radiocarbon Dates VIII. *Radiocarbon*, **17**, 226–238.

PENNINGTON, W. (1964). Pollen analyses from the deposits of six upland tarns in the Lake District. *Philosophical Transactions of the Royal Society of London* B, **248**, 205–244.

PENNINGTON, W. (1970). Vegetational history in the north-west of England: a regional synthesis. In: *Studies in the Vegetational History of the British Isles* (Ed. by D. Walker & R. G. West), pp. 41–80. Cambridge University Press, Cambridge.

PENNINGTON, W. (1981). Sediment composition in relation to the interpretation of pollen data. *Proceedings of the IVth International Palynological Conference, Lucknow (1976–77)*, **3**, 188–213.

PENNINGTON, W., HAWORTH, E. Y., BONNY, A. P. & LISHMAN, J. P. (1972). Lake sediments in northern Scotland. *Philosophical Transactions of the Royal Society of London* B, **264**, 191–294.

PIGOTT, C. D. & PIGOTT, M. E. (1963). Late glacial and post-glacial deposits at Malham Tarn, Yorkshire. *The New Phytologist*, **62**, 317–334.

PILCHER, J. R. (1969). Archaeology, palaeoecology, and ^{14}C dating of the Beaghmore stone circle site. *Ulster Journal of Archaeology*, **32**, 73–87.

PILCHER, J. R. & SMITH, A. G. (1979). Palaeoecological investigations at Ballynagilly, a Neolithic and Bronze Age site in County Tyrone, Northern Ireland. *Philosophical Transactions of the Royal Society of London* B, **286**, 345–369.

RANKINE, W. F., RANKINE, W. M. & DIMBLEBY, G. W. (1960). Further investigations at a Mesolithic site at Oakhanger, Selbourne, Hants. *Proceedings of the Prehistoric Society*, **26**, 246–302.

SIMMONS, I. G. (1964). Pollen diagrams from Dartmoor. *The New Phytologist*, **65**, 165–180.

SIMMONS, I. G. (1981). Sea-level in The Mesolithic. In: *The Environment in British Prehistory*, chap. 3. (Ed. by I. Simmons & M. J. Tooley), pp. 83–89. Duckworth, London.

SIMMONS, I. G., RAND, J. I. & CRABTREE, K. (1983). A further pollen analytical study of the Blacklane peat section on Dartmoor, England. *The New Phytologist*, **94**, 655–667.

SINGH, G. & SMITH, A. G. (1973). Post-glacial vegetational history and relative land- and sea-level changes in Lecale, Co. Down. *Proceedings of the Royal Irish Academy*, **73B**, 1–51.

SISSONS, J. B. & BROOKS, C. L. (1971). Dating of early post-glacial land and sea level changes in the western Forth Valley. *Nature, Physical Sciences*, **234**, 124–127.

SMITH, A. G. (1958). Pollen analytical investigations of the mire at Fallahogy Td., Co. Derry. *Proceedings of the Royal Irish Academy*, **59B**, 329–343.

SMITH, A. G. (1970). The influence of Mesolithic and Neolithic man on British vegetation. In: *Studies in the Vegetational History of the British Isles* (Ed. by D. Walker & R. G. West), pp. 81–96. Cambridge University Press, Cambridge.

SMITH, A. G. (1975). Neolithic and Bronze Age landscape changes in northern Ireland. *Council for British Archaeology: Research Report* No. **11**, 64–74.

SMITH, A. G. (1981). Palynology of a Mesolithic-Neolithic site in County Antrim, N. Ireland. *Proceedings of the IVth International Palynological Conference, Lucknow (1976–77)*, **3**, 248–257.

SMITH, A. G. & COLLINS, A. E. P. (1971). The stratigraphy, palynology and archaeology of diatomite deposits at Newferry, Co. Antrim, Northern Ireland. *Ulster Journal of Archaeology*, **34**, 3–35.

SMITH, A. G., GASKELL-BROWN, C., GODDARD, I. C., GODDARD, A., PEARSON, G. W. & DRESSER, P. Q. (1981). Archaeology and environmental history of a barrow at Pubble, Loughermore, Townland, County Londonderry. *Proceedings of the Royal Irish Academy*, **81C**, 29–66.

SMITH, D. E., MORRISON, J., JONES, R. L. & CULLINGFORD, R. A. (1980). Dating the main postglacial shoreline in the Montrose area, Scotland. In: *Timescales in Geomorphology* (Ed. by R. A. Cullingford, D. A. Davidson & J. Lewis), pp. 225–245. John Wiley & Sons Ltd, Chichester.

SMITH, A. G., PEARSON, G. W. & PILCHER, J. R. (1971). Belfast radiocarbon dates III. *Radiocarbon*, **13**, 103–125.

SMITH, A. G., PEARSON, G. W. & PILCHER, J. R. (1973a). Belfast radiocarbon dates V. *Radiocarbon*, **15**, 212–228.

SMITH, A. G., PEARSON, G. W. & PILCHER, J. R. (1973b). Belfast radiocarbon dates VI. *Radiocarbon*, **15**, 599–610.

SMITH, A. G. & PILCHER, J. R. (1973). Radiocarbon dates and vegetational history of the British Isles. *The New Phytologist*, **72**, 903–914.

SWITSUR, V. R. & JACOBI, R. M. (1975). Radiocarbon dates for the Pennine Mesolithic. *Nature*, 256, 32–34.

TALLIS, J. H. (1964). The pre-peat vegetation of the southern Pennines. *The New Phytologist*, **63**, 363–373.

TURNER, J., HEWETSON, V. P., HIBBERT, F. A., LOWRY, K. H. & CHAMBERS, C. (1973). The history of the vegetation and flora of Widdybank Fell and the Cow Green reservoir basin, Upper Teesdale. *Philosophical Transactions of the Royal Society of London B*, **269**, 327–408.

WALKER, D. (1956). A site at Stump Cross, near Grassington, Yorkshire, and the age of the Pennine microlithic industry. *Proceedings of the Prehistoric Society*, **22**, 23–8.

WEST, R. G. (1961). Interglacial and interstadial vegetation in England. *Proceedings of the Linnean Society of London*, **172**, 81–89.

WEST, R. G. (1980). Pleistocene forest history in East Anglia. *The New Phytologist*, **85**, 571–622.

WOODMAN, P. C. (1977). Recent excavations at Newferry, Co. Antrim. *Proceedings of the Prehistoric Society*, **43**, 155–200.

ANTHROPOGENIC CHANGES FROM NEOLITHIC THROUGH MEDIEVAL TIMES

By G. W. DIMBLEBY*

Late of Department of Human Environment, Institute of Archaeology, 31–34, Gordon Square, London WC1H, 0PY, UK

Summary

The long period covered by this paper probably saw the most drastic changes wrought by man on the plant cover of these islands in the whole of postglacial history. Much of our information about vegetation changes comes from those parts of the country where peats and mires are abundant, which leaves those areas which were the cradle of successive cultural groups largely unrepresented. Sources of environmental evidence are now emerging even from this unpromising background, an important element amongst which are man's own works. The evidence of changes wrought by man may come from non-botanical as well as botanical material. Pollen, charcoal, phytoliths, seeds, mosses, and other macroscopic plant remains may be extracted, but indirect evidence may also be obtained from land snails, insects, bones (including small vertebrates), as well as from preserved soils and land surfaces.

In the various environments considered a trend from forest through clearance to pastoral or arable agriculture can usually be demonstrated. These changes may ultimately result in site deterioration on a large scale, the nature of which will depend upon the ecological conditions and the practices at work. In extreme cases peat may form or soil erosion take place. On base-poor soils, soil acidification may become so intense that the land is no longer usable for traditional agriculture and is allowed to degenerate to moorland or heath. In early historic times much degraded land came into use again for hill-grazing, a process which further depauperated both the flora and the soils.

Key words: Anthropogenic changes, agriculture, environment, prehistoric vegetation.

Preamble

It might appear that the course of my career, through forestry to archaeology, can owe little to my initial training as a botanist since the undergraduate courses I took contained little specifically in these fields. On the contrary, I would want to say that it was the very breadth of that training which made it possible for me to follow into such fields when the opportunity offered. When I came to the Oxford Botany School in 1936, Tansley was in his last term as Sherardian Professor, and Roy Clapham was my personal tutor. I came up with some knowledge of the British flora and this was where my interest had lain up till then. But I was soon introduced to the concepts of ecology, which were new to me then, and have remained the basis of my work ever since. Not only was Clapham my tutor for my 3 years as an undergraduate, but he was my supervisor for my first effort in research. The topic he gave me was Soil Oxidation – Reduction Potentials, a far cry from my initial preoccupation with flora, but an illustration of the breadth of ecological study to which I was exposed under his guidance.

This breadth is reflected in this paper, for I shall not confine myself to botanical evidence alone, but also draw upon other elements of the ecosystem to find evidence about contemporary vegetation and the part man was playing.

* Present address: 34 House Lane, Sandridge, St. Albans, Herts., AL4 9EW.

Introduction

The brief for this paper was that it should be in the field of my research. To do this in the time available is made somewhat easier by the fact that although my research has extended over this period, its cover has been very uneven. My own work cannot serve as a basis for a uniform exposition over this period of time, so I must also draw on other work. This period probably saw greater changes brought about by man than at any other time in the history of these islands, not even excluding today. Our knowledge is still patchy but growing rapidly; this is still a research field, and we are not yet able to stand back and present a full picture. For this reason I propose to spend some time discussing the sources of evidence. Then I will try to make some assessment of how these have contributed to what we know about the different periods. I am not offering many new facts but the approach may be unfamiliar and perhaps helpful to those not used to using archaeological sources.

Sources of Evidence

Conventional pollen analysis

The pollen analysis of mires and lake deposits, has, of course, become the primary source of evidence about the changes which have taken place in the vegetation of Britain through the postglacial period. The evidence produced by this technique can be interpreted not only in terms of a sequence of changes brought about by climatic change but, by taking into consideration the known ecological requirements of certain groups of plants which turn up consistently, such as the so-called weeds and ruderals, it is possible to read into the evidence a role played by man in bringing about changes in the vegetation. The identification of seeds and other plant remains from the sample material provides confirmatory detail to the evidence of the pollen. Whilst this approach has produced such widespread and consistent evidence of the hand of man, it has suffered from two important limitations. In the first place, because of the climatic and hydrographical conditions across our islands, we have more evidence for the wetter and more acidic areas of the north and west than we have for the lowland areas of the midlands, south and east. It is true that more intensive application to such areas is now partially redressing the balance, but the fact remains that the potential for this technique is limited in those very areas which have been most intensively settled by man and which have seen the greatest demographic changes. Clearly the wetter areas have been affected by man, perhaps relatively more dramatically because the ecosystem there often tends to be fragile, but it is not possible to extrapolate from these areas to what was happening elsewhere.

The second limitation is the vexed question of interpretation and this applies wherever this technique is used. It is still to some extent a matter of judgment to distinguish local pollen from regional pollen (at least on isolated sites) and this has recently been high-lighted by Wilkins (1984) who has shown that on the Isle of Lewis trees were present at times when evidence from pollen diagrams suggested a complete lack of trees. For both these reasons, therefore, additional sources of data would be desirable, particularly where we are trying to see the influence of man in sharper perspective.

Archaeological sites and deposits

It is not surprising that where man's occupation has been the most extended and the most intensive, we find the greatest number of archaeological sites spread over the landscape. Every one such site is a fragment of the past which has survived at least in part to the present day, and may contain levels in deposits which can be made to yield indirect or even direct evidence of past ecological conditions. Such contexts of evidence come under the following headings: Buried old land surfaces; Waterlogged levels (e.g. deep pits, wells); Distributed deposits (e.g. hillwash, blown sand). Such contexts may be surprisingly rich in source material. There are certain basic principles which control this. Some materials, e.g. charcoal, are not subject to chemical or biological decay and may therefore be preserved wherever they occur. The biggest threat to the preservation of charcoal is fragmentation, which can be brought about by drying, freeze and thaw, and even waterlogging if the charcoal is not fully carbonized.

Some materials are subject to chemical destruction; e.g. bone or molluscan shells in any acid milieu. On the other hand, in calcareous deposits mineral accretion can lead to preservation; seeds can be preserved in an identifiable state by the accretion of calcareous or phosphatic deposits.

Generally, however, the preservation of purely organic materials is governed by the microbiological activity in the deposits. Anaerobiosis reduces decay, but anaerobiosis can be produced by conditions other than waterlogging (see below). Bacteriostasis may also be brought about by chemical factors, notably the presence of toxic metals (especially copper) and also by high salt concentrations (e.g. in old salt workings). By far the most important bacteriostatic factor, however, is acidity. It is this that makes it possible to extract pollen from acidic soil materials.

Finally it should be mentioned that soil profiles may be preserved on archaeological sites and these can be of direct value in interpreting early ecological conditions. Furthermore, they can produce evidence of soil accumulation, erosion, or disturbance by human agency such as ploughing.

Before turning to the application of this sort of approach to the question of vegetation change, it may be useful to specify the more important sources of ecological information which can be used in such research, and consider briefly the circumstances in which they can most usefully be studied. There are few sites which cannot be made to yield one or more of these sources if appropriate investigations are put in hand.

Pollen. As just indicated pollen may be very well preserved in acid soils (pH < 5.5). Nevertheless, in some circumstances it can be found even in calcareous soils, in which case its distribution may be of exceptional significance (Dimbleby & Evans, 1974). Many natural soils contain pollen throughout the soil profile, and this can give an indication of past history. It follows that if a pollen-rich soil is buried at a given period, its pollen profile may indicate the broad sequence of events preceding burial. Perhaps more valuable is the pollen spectrum of the actual surface of a buried soil, which should be precisely contemporary with burial.

There is abundant evidence that the pollen in soil is derived primarily from the vegetation on the site (contrast a peat bog) and from the immediate locality. Its value for regional interpretation is therefore limited. It has been shown that there can be major differences between pollen spectra from archaeological sites and those from points adjacent to the site. One reason seems to be that at all periods man has introduced pollen-rich material on to his sites, thus giving strong over-representation of such pollens (Dimbleby, 1984).

Charcoal. It has long been recognized that charcoal from prehistoric sites can be identified, but interpretation has often been a problem (e.g. Salisbury & Jane, 1940; Godwin & Tansley, 1941). The very indestructability of charcoal itself can be a problem (it is often re-distributed unwittingly). Its greatest value is when in conjunction with other evidence, e.g. pollen. Microscopic fragments of charred plant material may prove to be of quite different provenance from wood charcoal. For instance, charred pteridophyte tracheids have proved to be significant in chalk soils (Dimbleby & Evans, 1974).

Seeds. These occur most commonly in waterlogged deposits (e.g. Lambrick & Robinson, 1979), but as indicated above, they can occur in considerable numbers in other contexts if they become impregnated with calcareous or phosphatic deposits (Green, 1979). It should be recognized that, with seeds perhaps more than any other type of material, the value for environmental interpretation is very limited despite the fact that they may be identifiable to species level. The majority are brought on to the site, e.g. with harvested crops, or are voided by animals and man with other food materials. The commonest context of phosphate-coated seeds is in deposits with evidence of sewage or excretion.

Another way in which seed evidence has been preserved is as impressions in pottery, tufa, etc. Pottery impressions are commonly of cereal grains, and have been intensively studied, but other species may be represented, and have given useful information about contemporary weeds. The value for wider environmental significance is obviously very limited.

Phytoliths. The opaline silica inclusions from grasses may be found in large numbers, particularly in ash deposits. Like charcoal they are very resistant to destruction. Though they cannot be identified to species with any certainty, there are occasions when they can be of general environmental significance.

Mosses. Mosses turn up in a variety of archaeological contexts and are usually in a condition permitting identification to species. As some species have strong ecological preferences their presence may be particularly valuable. However, moss is a material which was useful to men, so that the possibility of selective introduction has to be borne in mind. This applied with particular force to urban sites.

Molluscs. Small land snails can be useful ecological indicators, and identification is possible with a high degree of precision (Evans, 1972). Shells of these snails may be preserved in great numbers in calcareous soils. To some extent, therefore, they serve the purpose which pollen plays in acidic soils. Interpretation, however, has to be different, if for no other reason than that dispersal is so different.

Coleoptera. Insect remains, mostly of beetles, can be very abundant in locally waterlogged sites such as pits or wells. Identification to species is often possible on quite fragmentary remains, and as there is a strong element of ecological preference in this group they can be of considerable value as indicator species. It should be noted, however, that the chitinous exoskeletons are relatively resistant *to decay and therefore may be carried out of their usual habitat, e.g. in the droppings of insectivorous birds.*

Bones. Bone material may form a very large bulk of biological remains particularly in neutral or alkaline sites. Within the period we are concerned with here most of it relates to domestic animals, but increasing attention is being devoted to the bones of smaller vertebrates, which may be more closely related to the external environment: they may also be susceptible, of course, to selective concentration through the agency of predatory animals such as birds of prey.

Soils. It will be more appropriate to illustrate the nature of this evidence in reference to particular cases below.

Survey of Vegetative Change

In Prehistoric Time

Neolithic

The onset of the neolithic period is a cultural event. It has been described in a compehensible and readable form by Piggott (1981). It may be recognized by certain marker species in the pollen sequence, but had the neolithic culture not been brought to these islands at that time, it must be assumed that these elements would not have become conspicuous. The possible exception to this is the Elm Decline; it is remarkable how many first neolithic dates cluster around the Elm Decline, though a recent paper by Edwards & Hirons (1984) shows that traces of cereal pollen have been found at a number of sites in pre-Elm Decline deposits. These apparently indicate arable agriculture, presumably by neolithic man. It is also to be noted that it itself is not dated to one precise point in time, but has been dated in different places over a period of several centuries. In some places more than one Elm Decline has been recognized. Whatever view one favours can be supported by some of the evidence, so I do not propose to discuss the subject further as I am not in a position to add any new facts. I will content myself with observing that the elm is a most useful tree to man, but when one recognizes that the neolithic incomers would have been faced for the most part with towering trees with their crowns in the canopy, it is not easy to see how they could have reduced its population so drastically with the resources available to them.

It is often assumed that the neolithic people came into a landscape of continuous forest, at least below a tree-line variously assessed as from 275 to 600 m in different parts of the Highland Zone, apart from localized areas such as the surroundings of lakes, areas of marsh, etc. This, of course, leaves out of consideration any possible effect which mesolithic man might already have had, and we know now that locally at any rate he did open up the forest. In places this led to soil acidification and even the establishment of heath. But we still cannot be sure of what his effect might have been over the country as a whole. Our assessment of the mesolithic population depends on the finding of microlithic flints; these turn up on upland sites and in other places where for one reason or another they have persisted, for example protected by a covering of peat, but it is not safe to argue from their absence, particularly in areas which have been subjected to centuries or millennia of agriculture with all the possibilities this gives of mass soil movement. There has been much speculation about the primary vegetation in the shallow soil of the Chalk and good pedological reasons given that chalk soils could never be deep soils. It is still possible, however, that there has been mass movement of the soil mantle following clearance of primary forest.

This is speculation, but the environmental investigation of archaeological sites is beginning to provide some hard facts. For instance, Evans (1972) has investigated the Mollusca of a number of neolithic buried soils on the Chalk. Typically he finds a sequence from open woodland, through clearance phase to grassland, and finally in many cases, to arable land (Fig. 1). At Avebury he identified a 'woodland' phase in the lowest layers of the soil. In this case I tried to get a pollen sequence but pollen was not present in countable quantity. In a number of buried neolithic soils Evans found what he called 'subsoil hollows' which contained a woodland fauna and led him to deduce that these hollows were old tree-root holes. So it seems that sizeable trees did grow on the Chalk even to the extent of forming woodland. If in one place, why not generally?

Using these techniques it has been shown that woodland regeneration could take place, e.g. at Ascott-under-Wychwood on the Cotswold dipslope. Here pollen was found and indicated lime wood. So we are still not able to say what the primary condition was on lowland calcareous soils if one can speak in those terms.

The hypothetical example of mesolithic clearance and resultant soil movement which I gave above is rendered unlikely by other evidence which is now accumulating. Whilst there has undoubtedly been massive soil movement on the Chalk, there is little evidence of it in neolithic times; one would expect clearance to have been more marked than in the Mesolithic. Maybe the sites, even so, were little more than open islands in a sea of forest.

It has been possible to obtain pollen spectra from a number of neolithic soils in Chalk, and they broadly support this picture, given that the evidence of the pollen must be seen as applying only to the immediate locality. Occasionally there is some suggestion of soil movement [e.g. Kilham long barrow (Manby, 1976)], but it is on a miniature scale.

Before leaving the Chalk in the neolithic period mention should be made of Silbury Hill, a middle to late neolithic monument (approx. 2500 BC) (Fig. 2). Various attempts have been made over the last century to determine its purpose, but its neolithic date, if not its purpose, was established by Professors Atkinson and Piggott, who in the 1960s drove a shaft to the centre at the old ground level. This proved invaluable for obtaining environmental evidence in the old land surface beneath the mound (Fig. 3). This work is still unpublished, so all I can do here is give a thumb-nail sketch of the main findings. The mound is built on an island of Clay-with-Flints, so the soil was acid enough to contain pollen. There was a central core of turves cut from the surrounding chalk soil, and these contained molluscs. Small-sized charcoal (hazel) was also recovered. This might have been expected, but what was quite unexpected was the preservation of uncharred organic remains, including the grass sward itself, mosses, seeds (Fig. 4), and even winged ants (indicating the time of year the construction took place). The broad interpretation was that the environment at this time was almost as open as now, but with a significantly greater representation of hazel in the pollen spectrum. What is perhaps even more important is that there in the midst of typical chalkland was preserved an assemblage of organic remains of a variety at least comparable with what might be found in many a waterlogged site. The explanation seems to be in the size and nature of the mound itself, which stands about 40 m high. Conditions were clearly anaerobic – there was blue vivianite in the buried soil – presumably owing to two main factors (a) that there was a large amount of organic

Anthropogenic vegetational changes

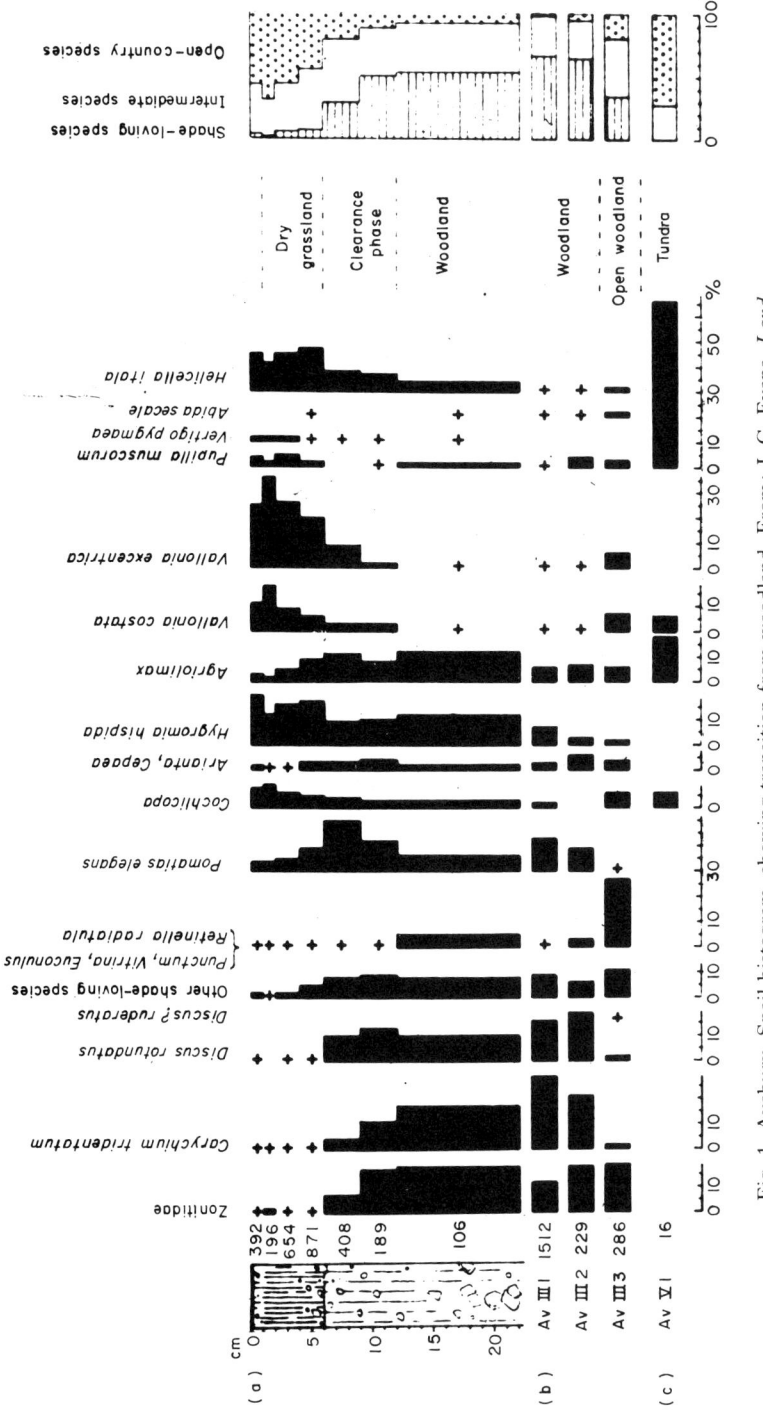

Fig. 1. Avebury. Snail histogram, showing transition from woodland. From: J. C. Evans, *Land Snails in Archaeology*. Seminar Press, 1972. (Copyright Seminar Press.)

Fig. 2. Silbury Hill set in the Marlborough Downs.

Fig. 3. Silbury Hill; old land surface under mound. Chalk turves and rubble overlying incipient stagnogley developed in loess layer on top of Clay-with-Flints.

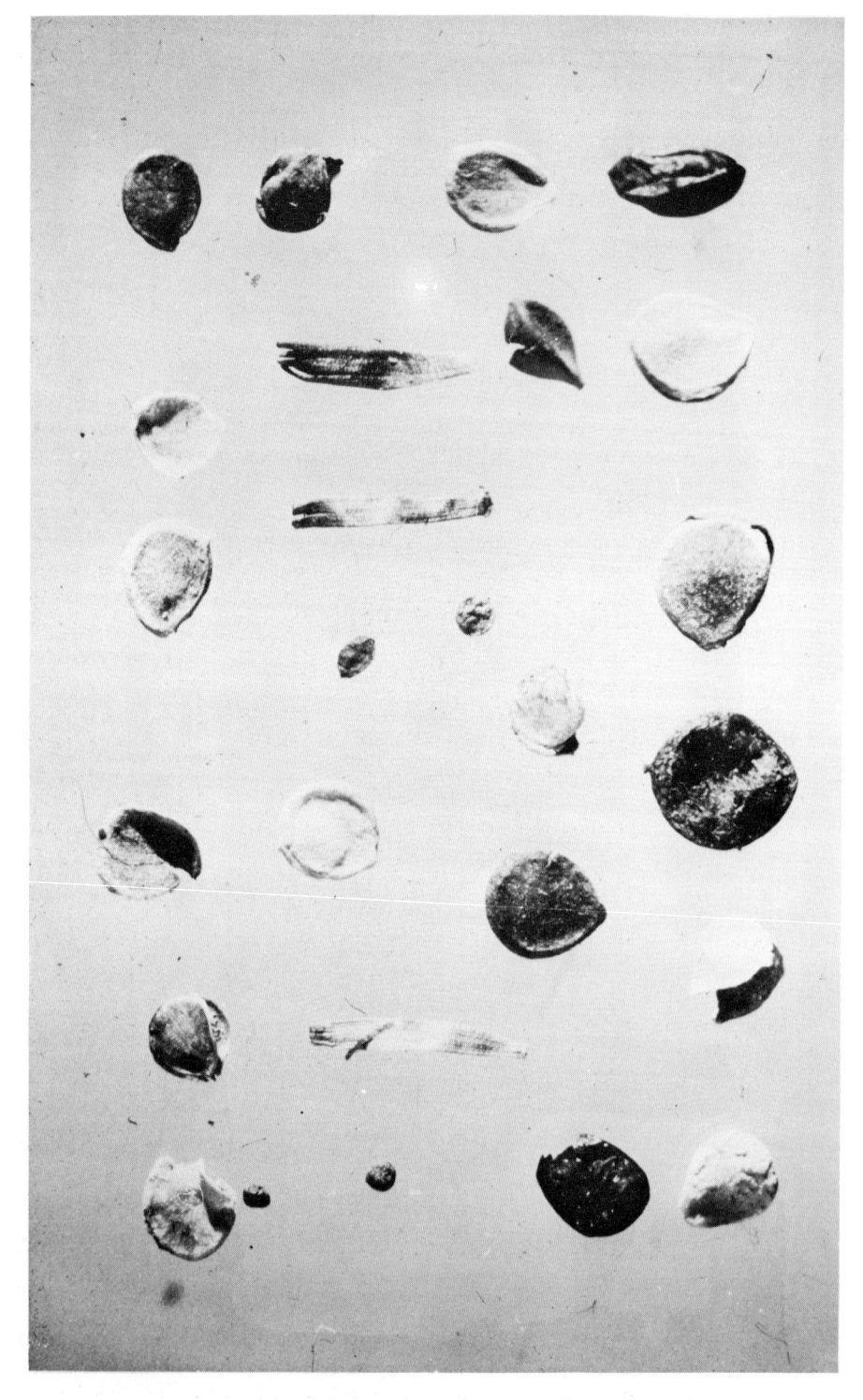

Fig. 4. Silbury Hill; uncharred seeds from old land surface.

matter in the turf core of the mound, which in decaying would use up any oxygen and (b) the large size of the mound covering this core which could have restricted a renewal of the oxygen level, so leading to oxygen deficiency.

Attention has been deliberately concentrated here on the calcareous soils because they are seldom treated in palaeoecological discussions, whereas they are central in terms of prehistoric man's interest. This is not to say other areas were avoided, though on the whole areas with soils of low base-exchange capacity were not used in this period. There were, however, other regions focused on by neolithic farmers, such as the Lake District, and Ireland. Indeed the dates of some of the Irish neolithic sites are among the earliest in Britain. Many of them are preserved under blanket peat, which even started to form during the neolithic period, though more generally in Britain its onset would be traced to the Bronze Age or even later periods. The fact that there is such a chronological range in the onset of blanket peat formation, and that at its base traces of pollen species associated with human activity may be found, suggests that early land use, in conjunction with climate, was a triggering factor. This introduces the concept of environmental degradation which will become more clear-cut in the later periods covered by this paper.

Whilst it seems likely that a simple factor, the removal of tree cover, would have instituted hydrological and pedological changes in the ecosystem, the actual operations of the prehistoric farmers would have added to this pressure. It is not yet possible to evaluate the selective importance of pastoral and arable agriculture; probably the former was dominant and in this connection it must be remembered that alien livestock, particularly the sheep, were introduced into Britain at this time. The native red deer ceased to be the important subsistence animal that it was in mesolithic times. There is evidence that neolithic people ranged widely beyond their settlement areas, perhaps hunting, but perhaps also following their animals.

At the same time, evidence of cultivation in the form of ard marks is increasingly being found on neolithic land surfaces preserved beneath earthworks (e.g. South Street long barrow), blown sand (Isle of Harris) and blanket peat (Ireland). [For a fuller treatment of this subject see Fowler (1983) and Mitchell (1976)]. Because of the disjunct distribution of suitable conditions for the preservation of such features, and probably the failure of earlier excavators to recognize them, it is only possible to say that arable agriculture was widespread, though perhaps on a small scale, in this period.

Finally, here it may be observed that the changes in the landscape were not all one-way. Pollen analysis at individual sites suggest that re-growth of woody species, particularly hazel (*Corylus*), did occur. Much would depend, no doubt, upon whether grazing pressure was maintained after clearance. The ard marks usually seem to represent a single episode of tilling which might suggest that plots were used once and then allowed to revert either to grass or even to woodland.

Bronze Age

I have dwelt at some length on the neolithic period, even though the direct evidence is often fragmentary, because it was at that time that many of the anthropogenic forces at work in the vegetation first made their appearance. As we come to the Bronze Age and later periods the actual processes are broadly the same, but their application develops. Fowler (1983) has given a very complete account of later prehistoric farming in Britain which I recommend as essential background to what follows.

Bronze Age occupation was very widespread, not only in areas such as Wessex that were centres of neolithic culture, but also in territory that had been neglected in the Neolithic. For instance, round barrows and cairns are a prominent feature of many moorland areas, whereas long barrows (Neolithic) are virtually absent from such areas.

Again it is difficult to assess the relative importance of pastoral and arable farming, but it is clear that whilst these practices continued on land which had long been farmed, the frontiers of farming were pushed outwards. This is particularly well exemplified by the pattern of reaves (low banks) which cover extensive areas of Dartmoor, and which apparently represent boundaries in a system of pastoral farm or ranching. Similar features can also be seen on Salisbury Plain, an area settled in the Neolithic. It seems that this Bronze Age system was superimposed on the old landscape. In other parts of the so-called Highland Zone, field systems, apparently connected with arable farming, can be traced on aerial photographs of land which is now moorland. I have myself picked up Bronze Age saddle querns on moorland newly ploughed by the Forestry Commission. These are too heavy to be casually distributed (I knew one archaeologist who ruptured himself carrying one home), so they were presumably in use on the site. This pushing out of the frontiers of farming would have had two main effects on the landscape. In the first place the area of primeval vegetation, whatever that was, would have been reduced, and secondly, the removal of the tree cover in areas previously untouched would have initiated changes in the ecosystem, particularly on the soil.

It would probably be futile at the moment to attempt to give an overall picture of what Bronze Age Britain looked like, so I prefer to concentrate on some of the trends. Our conventional pollen analyses show clearly the change from woodland to open country to be very active in this period, and on basic soils the mollusc sequences reflect the same processes. I have reported elsewhere (e.g. Dimbleby, 1978a) the changes which can be traced on one piece of what is now moorland (Great Ayton) by comparing pollen spectra under earthworks of three ages, late Neolithic, Bronze Age and Iron Age (Fig. 5). The late neolithic site was set in a clearing in deciduous forest; by the Bronze Age the area was more open with mainly pastoral farming, but with a strong background of hazel, whereas by the Iron Age the surroundings were as open as now with pastoral farming predominant, and the hazel element much reduced. It was not until after the Iron Age that heather (*Calluna*) took over.

I cannot here go into detail about trends of soil change which are revealed by studies of the buried soils under such earthworks (see Dimbleby, 1978b); the general pattern is that on soils of low base status deforestation is followed by acidification and continued use eventually leads to some form of podzolization. Podzolization can precede the eventual spread of heather.

In the last section I mentioned the onset of peat formation, a process which continued through the Bronze Age. Peat formation seems to be the form of site degradation associated with areas of higher rainfall, whereas podzolization (contrary to some text books) replaces it in drier areas.

This progressive acidification of soils on acid rocks can sometimes be traced back, if not to the forest condition, at least to a soil which was originally a brown soil, carrying pollen evidence of farming in an environment of hazel scrub. The Bronze Age farmers clearing forest on such land would have encountered brown soils of relative fertility (compared with today) but a fertility which proved transient.

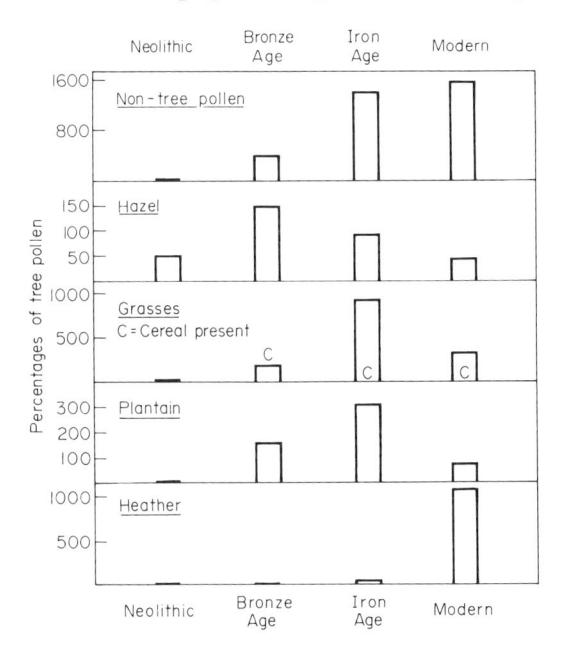

Fig. 5. Great Ayton Moor, North Yorks. Pollen analyses of buried surfaces.

The basic soils apparently were able to carry prehistoric farming through the Neolithic and the Bronze Age and yet remain productive. That they too could deteriorate will become apparent in the next section. Before leaving this period, however, reference must be made to another site in the Chalk which, as at Silbury Hill, but for very different reasons, preserved a remarkable assemblage of organic materials of various kinds, reflecting the local environment. This was a site at Wilsford, near Stonehenge. This was thought to be a pond barrow, but initial excavation revealed no natural subsoil even at 6 m depth. A major excavation was therefore instituted, which showed that the so-called pond barrow was in fact of ring of waste thrown up from a narrow shaft over 30 m deep. It was waterlogged at the bottom and produced artifactual remains such as bast rope and an alder wood stave bucket, but also natural organic materials including mosses (still green), twigs, cereal straw, abundant insect remains (Osborne, 1969), and pollen. In total this evidence indicated open surroundings with both arable and pastoral farming; the mosses indicated that there was short-grass downland in the vicinity, a conclusion confirmed by the pollen analysis which included *Poterium* and *Helianthemum*. This work, like Silbury Hill, has not yet been published definitively.

Iron Age

Reference again should be made to Fowler (1983). Perhaps the most significant development in this period was the introduction of iron farming tools, particularly the iron plough, which enabled land to be ploughed which the wooden (or even iron-shod) ard could not cope with. It was therefore a period of intensified arable agriculture. Much of this continued on soils which had long carried arable farming; the soils which had been broken in acid parent materials during the Bronze Age had, as we have seen, undergone degradation, to the extent of going out of production in many cases. Obviously it would not pay to re-cultivate such areas (though I am leaving out here any discussion of knowledge of the principles of manuring).

This concentration of arable practices with more effective tools, perhaps combined with a tendency to reduced periods of fallow, had one major result – soil movement. This manifested itself in two main ways. First, there is the formation of the so-called Celtic fields, appearing on steep slopes, having been preserved by the development of undisturbed grassland over them. They are marked out by low banks or lynchets. There is good evidence that the position of the lynchets was pre-determined, perhaps by a wall or hedge, but the build-up of soil along these lines could have been due to the action of gravity and probably rainwater on soils loosened by repeated tillage. The continual removal of the topsoil would expose material with progressively less crumb structure and so possibly accelerate the process.

Celtic fields have been known and discussed for a long time, but the other form of soil movement is mass movement down hill. It has been argued (Godwin, 1976b) that erosion cannot occur on porous chalk soils. Recent investigations into the fill of dry valleys (Bell, 1983), however, have shown this fill to be several metres deep and to contain archaeological artifacts of successive periods, going back at least as far as the Bronze Age. It is irrefutable that colluviation has taken place at various times, removing the earlier soils containing datable artifacts. The intensification of ploughing in the Iron Age seems to have been the main initiator of this process which continued right up to and even beyond medieval times. Celtic fields were preserved because cultivation ceased and they became grassed over, but this process of colluviation will have resulted from the continued cultivation of the dry valley sides. Where grassland exists on these slopes today, therefore, it must be much more recent in origin than that preserving Celtic fields.

In Historical Time

Roman

Unlike most history books, works on environmental archaeology tend to get less detailed as they come towards historical time, and this certainly applies to this paper. One reason is that the sources of information change. The written word becomes available and tends to replace the evidence of field work. Furthermore, the development of towns and ports shifts the archaeologist's interest from the countryside to more urban situations.

These constraints certainly begin to operate in the Roman period and, though Iron Age practices continued to provide rural sites, dedicated Romanists find greater satisfaction in the richer urban investigations. Nevertheless, it seems to be generally accepted that the Roman's main effect on the landscape, and therefore on the vegetation, was to intensify arable production within the existing pattern, whilst also setting up many villa farmsteads, and at the same time opening up communication between areas previously isolated from each other. It is surprising that with such a sudden increase in population, and the fact that one of the reasons for invasion was to tap the cereal-growing potential of Britain, there is no stronger evidence of environmental stress. Pollen analyses of this period uniformly suggest that in the south of England the landscape was very arable-dominated, though in the north and west, much more woodland remained. However, we need more critical evidence for this period. It is often assumed that pressure on the land means pressure on the remaining woodland resources, for timber or for usable land. Whilst this is generally true, the Romans also needed a renewable supply of fuel.

Hundreds of villas are known; not only would each have its own often extensive field system, but each one typically had its own bath-house. It has been calculated that for one small bath unit it would be necessary to have approximately 20 ha of coppice woodland permanently in production. It is an interesting question to what extent coppice woodland would be recognizable in pollen analyses. The hazel element in contemporary spectra would seem to warrant particular attention.

Recent archaeological work in Upper Thames Valley (Lambrick & Robinson, 1979), using pollen analysis, seed identification and insect analysis, is producing new insight into the agricultural use of the terraces and the floodplain grassland, and the way this changed from the Iron Age to Roman times. Such work is not yet extensive enough over the country to form the basis of wider interpretation of our vegetation and flora, but it holds great potential. A useful collection of papers dealing with the environment in the Iron Age to the Anglo-Saxon period is found in Jones & Dimbleby (1981).

Anglo-Saxon

If the Roman period is poorly represented in the environmental literature, the same is even more true of the Dark Ages. Once again archaeological interest centres on urban and seaport sites, with a few exceptions such as Sutton Hoo. However interesting these may be in themselves, they do not advance our knowledge of British vegetation to any significant extent. The historical record, such as it is, is now the main source.

One general point that can be made is that from the end of the Iron Age there was a tendency to abandon some of the agricultural land on the uplands and their slopes and move into the valleys and lower lying ground. This has been interpreted as the result of loss of fertility on the traditional sites, but equally it might be due to the fact that more substantial implements of cultivation made it possible to exploit heavier soils that could not previously be worked.

The development of the mould-board plough, probably a Roman innovation, made possible ridge and furrow ploughing on the heavier soils, which became such a feature of our landscape from this time on.

Medieval

In this period historical record is our main source of information about British vegetation. Pollen analyses do fill in details; for example, Godwin's (1967a) work at Old Buckenham Mere in which he demonstrated the growing importance of hemp (*Cannabis*) as a crop. Urban sites increasingly occupy the attention of archaeologists, often because re-development of town centres gives an unrepeatable opportunity for study. Environmental studies are regularly being carried out alongside this work and are producing questions which reflect back to the activities in the countryside. Nevertheless, this remains a very limited contribution at present, and much more work is needed to put flesh on the skeleton provided by documentary records. Here it is possible to do no more than mention the evidence of population changes in this period: on the one hand the development of such features as strip lynchets, apparently to give more land for cultivation, and on the other hand the many villages which were deserted in the 14th century following visitations of the plague and possibly changes of food production enforced by climatic change.

This period saw the peak of the monasteries, which controlled vast areas of land. The Cistercians in particular, setting their houses in remote places, brought old degraded lands into new use. Heather had become the dominant plant in many upland areas; others were under grass of various kinds. All formed a basis for an explosion of sheep farming by the monasteries, starting a pattern which by today is regarded as traditional. The sheep grazing not only destroys woodland and effectively prevents its regeneration but it exerts a selective pressure against certain species leading to a species-poor vegetation. It is interesting to observe the richness of the flora on crags and places inaccessible to the sheep. Such places are refuges not only for rare species but also for woodland relicts.

Conclusion

As indicated at the outset, this paper had to be concerned with vegetation rather than flora, since this has been the direction of my own research. To balance this, however, mention should be made of the increasing body of knowledge of the introduction of new species into our flora by man (Godwin, 1975), which has taken place at all stages of the period covered by this paper.

Finally, it is appropriate to draw attention to the fact that we are fortunate in still possessing small areas of vegetation which, whilst not unaffected by man, have escaped the more destructive influences of agricultural development. Areas such as the New Forest, and certain small woodlands in eastern England (Rackham, 1980) are invaluable ecological survivors and should be treasured.

References

Bell, M. (1983). Valley sediments as evidence of prehistoric land-use on the south Downs. *Proceedings of the Prehistoric Society*, **49**, 119–150.

Dimbleby, G. W. (1978a). *Plants and Archaeology*. 2nd Edn. John Baker, London.

Dimbleby, G. W. (1978b). Changes in ecosystems through forest clearance. In: *Conservation and Agriculture* (Ed. by J. G. Hawkes), pp. 3–16. Duckworth, London.

Dimbleby, G. W. (1984). *Palynology of Archaeological Sites*. Academic Press, London. (In press).

Dimbleby, G. W. & Evans, J. G. (1974). Pollen and land-snail analysis of calcareous soils. *Journal of Archaeological Science*, **1**, 117–133.

Edwards, K. J. & Hirons, K. R. (1984). Cereal pollen grains in pre-elm decline deposits: implications for the earliest agriculture in Britain and Ireland. *Journal of Archaeological Science*, **11**, 71–80.

Evans, J. G. (1972). *Land Snails in Archaeology*. Seminar Press. London.

Fowler, P. J. (1983). *The Farming of Prehistoric Britain*. University Press, Cambridge.

Godwin, H. (1967a). The ancient cultivation of hemp. *Antiquity*, **41**, 42–49.

Godwin, H. (1976b). Strip lynchets and soil erosion. *Antiquity*, **41**, 66–67.

Godwin, H. (1975). *History of the British Flora*. 2nd Edn. University Press, Cambridge.

Godwin, H. & Tansley, A. G. (1941). Prehistoric charcoals as evidence of former vegetation, soil and climate. *Journal of Ecology*, **29**, 117–126.

Green, F. J. (1979). Phosphatic mineralization of seeds from archaeological sites. *Journal of Archaeological Science*, **6**, 279–284.

Jones, M. & Dimbleby, G. W. (Eds.) (1981). *The Environment of Man: the Iron Age to the Anglo-Saxon period*. British Archaeological Reports (British Series) vol. 87, Oxford.

Lambrick, G. & Robinson, M. (1979). Iron Age and Roman riverside settlements at Farmoor, Oxfordshire. Council for British Archaeology Research Report 32, London.

Manby, T. G. (1976). The excavation of the Kilham long barrow, East Riding of Yorkshire. *Proceedings of the Prehistoric Society*, **42**, 111–159.

Mitchell, F. (1976). *The Irish Landscape*. Collins, London.

Osborne, P. J. (1969). An insect fauna of late Bronze Age data from Wilsford, Wiltshire. *Journal of Animal Ecology*, **38**, 555–556.

Piggott, S. (Ed.) (1981). *The Agrarian History of England and Wales*; 1i, Prehistory. University Press, Cambridge.

Rackham, O. (1980). *Ancient Woodland*. Edward Arnold, London.

Salisbury, E. J. & Jane, F. W. (1940). Charcoals from Maiden Castle and their significance in relation to the vegetation and climatic conditions in prehistoric times. *Journal of Ecology*, **28**, 210–325.

Wilkins, D. A. (1984). The Flandrian woods of Lewis (Scotland). *Journal of Ecology*, **72**, 251–258.

POST-MEDIEVAL AND RECENT CHANGES IN BRITISH VEGETATION: THE CULMINATION OF HUMAN INFLUENCE

By D. A. RATCLIFFE

Nature Conservancy Council, Godwin House, George Street, Huntingdon, PE18 6 BU, UK

Summary

Although measurable climatic fluctuation in temperature and wetness has occurred between A.D. 1700 and 1984, lack of earlier records and the complicating effects of human influence militate against detection of vegetational response in Britain. Vegetational change during this period is thus regarded as almost entirely the result of anthropogenic factors.

Agriculture, now affecting c. 80% of Britain's surface, is overwhelmingly the most important such factor. Original vegetation had been mostly destroyed or heavily modified by A.D. 1700, notably through removal of the former forest cover and draining of lowland wetlands. The last 300 years have seen continuing development of farming and other land uses, to the point where very little natural vegetation now remains, and the main extent of semi-natural types is located in the uplands.

Many distinctive semi-natural vegetation types actually created by man over several centuries of low input–low output management have decreased greatly since A.D. 1700, and rate of loss has even accelerated since 1940–1950. Only 5% of permanent lowland grasslands now remain agriculturally unimproved neutral grassland and only an estimated 20% of the extent of pre-1940 calcareous grassland is now assignable to that category floristically. Lowland acidic heath has declined by 78% in area since A.D. 1830, and by 40% of the 1950 level. Lowland rich fens are reduced to scattered remnants, the largest expanse (the East Anglian Fenland) having decreased from 3380 km^2 in AD 1630 to 10 km^2 in 1984.

Forest cover was reduced to 5·4% by 1924, but has increased to 9·4%, largely through planting of alien conifers in upland districts. Since 1930 an overall 46% of ancient semi-natural broadleaved woodland has been lost, largely by conversion to conifer plantation but also through grubbing out. An estimated 300000 ha of ancient semi-natural woodland remains, representing 14% of existing woodland and 1·3% of Britain's land surface. The total extent of hedges in England and Wales was reduced by an estimated 140000 miles (28%) between 1946–1947 and 1974.

Urban–industrial growth, including roads, railways, water-use, mineral extraction and energy generation, has directly obliterated large areas of vegetation and caused chemical pollution. Waste ground associated with these activities often develops semi-natural plant communities through spontaneous succession, and may become floristically rich. Pollution affects especially fresh water through both toxic residues and eutrophication from nutrient enrichment (including run-off of agricultural fertilizers). Atmospheric pollution causes acid deposition especially damaging locally to the lichen flora, and induces biological changes in lakes and rivers in districts of acidic rocks.

Recreational pressures have caused increasing disturbance to vegetation locally, and less directly through working of *Sphagnum* bogs for horticultural moss litter and surface mining of limestone pavements for rockery stone. Collecting of plants as a hobby has also brought some species, notably rare ferns, orchids and 'alpines' to dangerously low population levels, or even local extinction.

The British flora has been profoundly modified by human influence. Only 19 vascular species are known to have become extinct, but many have declined greatly during the last 300 years and, for species in some habitats, especially semi-natural grasslands, rate of decrease has accelerated since 1940. Of 317 nationally rare species 117 (37%) have declined by at least 33% since 1930. Human agency has also introduced to Britain at least 239 plant species which now grow wild, and extended the distribution of at least 363 others. Some are still increasing but others, notably weeds of arable land, are decreasing again.

Conservation measures include setting up nature reserves to contain adequate representation of all semi-natural and natural vegetation types, and of populations of all native plant species. The aim is to give protected area status to *c.* 10% of Britain's surface through appropriate control and management designed to maintain or enhance the nature conservation interest. At present *c.* 1% is managed as nature reserve of one or other category and another 6% has been designated as Sites of Special Scientific Interest. Wild plants are now legally protected against unauthorized picking and uprooting, and 61 rare species are protected by special measures. Wider environmental conservation measures to reduce damaging human impact on vegetation and flora include inputs to and by local authorities and their planning system, persuasion of landowners and resource managers, and pollution control; supported by an educational and publicity programme, and by relevant research and information.

Key words: Post-medieval vegetational changes, recent vegetational changes, British vegetation, human influence.

INTRODUCTION

This most recent period in the history of British vegetation is but a small fraction of the late Quaternary epoch. The last 300 years are regarded as a segment of the Sub-Atlantic Period which began 2500 to 2800 years ago (Godwin, 1975). While it is necessary first to consider the nature of recent climatic shifts and their possible effects, this is too short a time for any large-scale change in climate to have occurred. Vegetational changes during this post-medieval period are largely attributable to the intense human activity which has been the dominant ecological influence. The present paper is accordingly concerned largely with an appraisal of the different aspects of this most recent phase of man's impact, in causing change in both distribution and composition of vegetation and flora.

CLIMATIC FLUCTUATION A.D. 1700 TO 1984

Lamb (1982) has shown that the most marked climatic fluctuations during the last 300 years have been in temperature. The beginning of this period coincided with the approximate end of the 'Little Ice Age' (*c.* A.D. 1550 to 1730). A trend of rising mean temperature then continued, albeit in a somewhat step-wise fashion, up to about 1940 to 1950. The total increase is estimated as 1 °C in annual mean, and the accurately recorded increase for 1900 to 1940 was 0·5 °C, marked by retreat of Northern Hemisphere glaciers and Polar pack-ice. Since 1940 to 1950 there has been another downturn in mean temperatures, and by 1980 a decrease of 0·3 °C was registered.

These temperature fluctuations could theoretically be expected to have measurable effects on plant and animal distribution. Ford (1982) has examined the evidence for such biological response to recent climatic change, and found that animals provide the best examples. Migratory species tend to be responsive, but invertebrates are especially sensitive and provide some well-documented correlations between temperature change (both as weather extremes and longer-term trends) and shift in distribution limits. By contrast, plants appear to be more buffered against temperature change, at least in their vegetative performance, and it is reproduction and establishment which are most vulnerable. The documentary evidence for recent botanical response to temperature change concerns especially the cultivation of fruit, such as grapes and peaches, which belong essentially to warmer climates than that of Britain (Lamb, 1982).

Both thermophilous southern and cold-adapted northern plant species at the limits of their range in Britain should, in theory, be sensitive to temperature shifts and unusual extremes. Small colonies of filmy ferns (two of *Hymenophyllum tunbrigense*, one of *Trichomanes speciosum*) in Cumbria and Dumfriesshire died out in the unusually severe winter of 1962 to 1963. These might seem to be instances of the cutting back of range by elimination from localities at the limits of distribution. The populations in question were, however, depauperate remnants in habitats made marginal by human forest modification, and healthy colonies of these species in the same district elsewhere survive unchanged. Many populations of *H. tunbrigense*, even in south-west Ireland, showed marked browning by frost during this winter, but made general recovery subsequently. Some populations of *Taxus baccata* in Cumbria were also severely browned, but later recovered fully. *Ulex europaeus* was extensively killed in parts of western Ireland by the severe frosts of 1981 to 1982, but this species regenerates freely from buried seed, and is unlikely to show any permanent depletion. Die-back of quite large patches of *Calluna vulgaris* through frosting is fairly common locally, but regeneration of the patches usually follows.

The northern maritime *Mertensia maritima* showed a marked loss of southern British localities between 1930 and 1960 (Perring & Walters, 1962). This species grows on unstable shingle and sand, where virtually all colonies are vulnerable to storm destruction, but failure to re-establish has been mainly in southern localities. The Mediterranean *Otanthus maritimus*, which occurs in similar habitats, has disappeared from all its former 20 localities in southern England and Wales, apparently for the converse reason, of retreat during an increasingly colder period (Perring & Farrell, 1977). Good (1936) correlated the spread of the continental-southern orchid *Himantoglossum hircinum* in south-east England during 1900 to 1933 with increasing winter and spring temperatures and overall increase in winter rainfall (i.e. increase in oceanicity). The subsequent decline of this species matches the reversal of this climatic trend.

Recent trends in wetness of climate are, however, more uncertain than those of temperature. Lamb (1982) has shown a marked fluctuation in frequency of south-westerly surface winds (which tend to give rainfall) in England from A.D. 1700 to the present. There were peaks in frequency around 1730, 1870 and 1922. Westerly winds over the British Isles increased from an average 85 days per year in the 1880s to 110 in the 1920s and then decreased again to 70 in the 1970s. Ford (1982) has commented on the tendency for fungal diseases of plants to flourish during especially humid seasons. In Ireland, the warm wet summers of 1845 to 1848 provided ideal conditions for the spread of the recently introduced *Phytophthora infestans* which caused the great potato famines. Ford suggests, conversely, that the hot, dry summers of 1975 and 1976 accelerated the spread of the introduced Dutch elm disease by encouraging activity and dispersal of the bark beetle *Scolytus scolytus*, vector of the causative fungus *Ceratocystis ulmi*.

Variations in climatic wetness from one year to another are well known as the cause of differences in yield of various crops and, to some extent, in seed production by native species, e.g. *Fagus sylvatica*, though these are again effects mainly on reproduction. The extreme summer drought of 1976 caused widespread death of plants through desiccation, and indirect effects such as increase in beech (*F. sylvatica*) bark disease and catastrophic heath/moorland fires. The species affected were, however, mostly common and widespread, and no appreciable change in abundance or distribution resulted, though local adjustments in competitive balance between species may have occured. Wind is another important element of an oceanic climate, and variations in windiness affect the dispersal capacity of many plant species, as well as causing differences in frequency and amount of storm damage to various types of vegetation.

These annual variations of climate clearly create something of a flux in plant performance and competitive balance between species. There is, however, remarkably little evidence of any clearly defined *trends* of change in plant abundance and distribution in response to the large-scale climatic fluctuations which have occurred since the Middle Ages. The geographical pattern of retreat of several plant species is only suggestive in its broad correlation with climatic change, and there is a dearth of evidence for geographical or altitudinal shift in limits of vegetation types or community dominants. Perhaps the period over which critical botanical and vegetational records have been available is simply too short. The temperature shift since c. 1700 is at least half of that estimated to have occurred during the whole of the time between the post-glacial thermal maximum and the present (2·0 °C – Godwin, 1975); seemingly large enough to have produced measurable effects on natural vegetation. Perhaps, as well as climatic extremes, the duration or frequency of a climatic peak or trough is important in overcoming the vegetative lag-effect in biogeographical response of perennial plants. There may also be threshold effects after which such change is rapid. Annuals would seem to be less cushioned against climatic change than perennials, but the known longevity of buried seed suggests that plant survival over unfavourable periods is a complex matter. The different components of climate also interact considerably, so that even broad correlations between biological phenomena and measured change in climatic parameters can be difficult to establish.

An over-riding factor is, however, the confusing effect of interacting human influence. Most of the sparse British evidence relates to introduced species and crop plants or situations modified by man. The pervasive anthropogenic destruction and disturbance affecting natural vegetation, the changes in abundance and distribution of so many species, and the restrictions to spread of other species, are likely to have limited both the botanical response to climatic change and the possibility of identifying such change as may actually have occurred. In regions of north-west Europe where human influence has been negligible there is good evidence of such change, in response to the recent period of warming climate. In Finland, Erkamo (1952) reported a northwards extension of the distribution limits of *Fraxinus excelsior*, *Acer platanoides*, *Tilia cordata*, *Typha latifolia*, *Typha angustifolia*, *Hypericum hirsutum* and *Ricciocarpus natans* during 1925–1950. Aas (1969) found that over the period 1918 to 1968 the average upper birch (*Betula pubescens*) timber line had risen by more than 40 m in the eastern highlands of southern Norway. In the southern Swedish Scandes, Kullman (1979, 1981), found evidence of a rise in upper limits of both *B. pubescens* and Scots pine (*Pinus sylvestris*) by around 30 to 40 m during 1915–1975. A natural timber line has virtually ceased to exist in Britain so that there is little possibility of detecting such trends. Slow change in size and frequency of populations of species at the limits of geographical range is likely to be the most readily detectable response to climatic shift here.

The Anthropogenic Character of British Vegetation in the Twentieth Century

The modification of the Sub-Atlantic climatic and edaphic range of vegetation by human influence has intensified during the last 300 years. The evidence is largely indirect, since nobody in A.D. 1700 was concerned about base-lines for botanical monitoring. Indeed, the period 1700 to 1900 was the great age of botanical description and taxonomy, and then of geographical recording in county floras, while description and classification of vegetation belong to the twentieth century. Even now, we do not yet have a standard taxonomy of vegetation types in Britain. The system assembled by Tansley (1939) is still widely used as the framework of reference, though wider survey has filled most of the gaps in his geographical coverage and there has been increasing description of British vegetation by phytosociological methods. The Nature Conservancy Council has contracted Professor C. D. Pigott and Dr J. Rodwell, with a team of collaborators, to produce a countrywide phytosociological classification of vegetation to be published in 1986. This will allow more precise comparisons with the vegetation of continental Europe and Ireland described by such methods.

Knowledge of vegetational change during 1700 to 1984 depends largely on the methods of historical ecology, using maps and documentary records. Some pollen, diatom and peat stratigraphic evidence is available and, during the last 100 years, the direct record of photographs, detailed vegetation maps, permanent quadrats, written accounts and verbal testimony has been increasingly available to document recent change. Inferential evidence is also important, and drawn especially from comparisons between analogous situations differing mainly in level of human impact, and from the backward extrapolation of short-term observations of actual change. For changes in distribution and abundance of species, the older Watsonian vice-county and variable local flora approaches have been replaced by the standard system of conventional grid mapping advocated in 1950 by Professor A. R. Clapham as a technique for monitoring change in the British flora. This system has since been adopted with great success for vascular plants in the pilot *Atlas of the British Flora* (Perring & Walters, 1962), and subsequently for most other major taxonomic groups of plants and animals in Britain. It has become virtually *de rigueur* as a recording method in modern county floras.

The culminating effect of a human population of 54 millions on an island measuring only 230000 km^2 has been to leave very little truly natural vegetation, i.e. that unmodified by man's activity. Natural plant communities are confined to inaccessible situations such as cliff faces and ledges, high mountain summits in the Scottish Highlands, some rivers and lakes, some coastal habitats (especially within the intertidal zone), and small fragments of various types elsewhere. There is nevertheless a considerable extent of vegetation of the kind Tansley (1939) termed 'semi-natural', i.e. composed of indigenous species and with a structure approximating to that of natural types. This is especially extensive in the uplands of the north and west, but it has in recent years decreased considerably in the lowlands, where rural land use has intensified and the built environment has spread. Increasingly there is vegetation of artificial character, composed of non-native species and occupying man-made habitats.

Man is able to create almost whatever vegetation he wishes, but, while gardens are scarcely an appropriate subject for inclusion here, human intervention is a continuum between totally deliberate and completely incidental, and it is difficult to draw a separating line. The proportion of the British flora owing its presence here to man is now quite considerable, and has to be regarded as an integral part of our vegetation (see below).

The Development of Agriculture

The recent history of British vegetation is dominated by the story of advancing agriculture, as the last phase in a continuous development spread over several millennia. Human use of the land to produce food has increasingly replaced or modified remaining original vegetation; not at a steady rate, but in irregular stepwise fashion, hesitating now and then or here and there before moving inexorably onwards again.

1700 to 1940

Farming was well developed as our economic mainstay by the end of the Middle Ages. It has involved the clearance of forest during the whole period of human occupation as detailed by Professor Dimbleby (pp. 57–72). By A.D. 1700 the bulk of our former tree cover had been replaced by farmed land, either of arable cropland or of grassland for pasturage of domesticated livestock. The other major source of farmland was from the draining of lowland wetlands – the shallow lakes, fens, marshes and raised bogs which occupied the flood plains of sluggish rivers and ancient marine sediments. It has also been found that when estuarine salt marshes were sealed off from the sea by embankments, they lost their salinity and could be converted to valuable grazing marsh. New saltmarsh began to form on the tidal side of the sea walls, and so there was an actual expansion of land area.

The original pattern of agriculture had developed with an ecological basis. The areas and habitats which were first settled and cultivated tended to be those where conditions were most favourable; dry and warm, with fertile, well-drained soils and a native vegetation not too difficult to clear. This pattern became reinforced during the modern period. Cereal growing was mainly in the most continental lowlands, with low rainfall and warm summers; while permanent pasture prevailed in the west, where heavier rainfall favoured grasses. There were, however, two main types of grassland. On deep, fertile loam soils, with pH 6·0 to 7·0 and good nutrient balance, were grass and forb communities with high biomass production – the class which Tansley (1939) termed 'neutral grasslands.' Some were on brown earths which formerly carried woodland, but others were on peaty or alluvial soils derived from fens or reclaimed saltings. They were all managed by grazing with sheep, goats, cattle or horses throughout the year, but on some the grazing was lifted during the spring and early summer to allow the sward to grow tall for cutting as hay.

The other main grassland type was on the outcrops of strongly calcareous rock which occur mainly in England: the Cretaceous chalk and the Jurassic, Magnesian and Carboniferous limestones. These parent materials produce soils varying from thin rendzinas to deeper brown earths, with pH usually in the range 7·0 to 8·0 and often containing free $CaCO_3$ but with a tendency to deficiency in some major plant nutrients, notably nitrogen, phosphorus and potassium. Their grasslands are characterized by a high proportion of markedly calcicolous herbs, but when ungrazed they produce a shorter and more tussocky, uneven growth with rather lower biomass than the previous type. These calcareous grasslands were used as pasturage too, especially for sheep, and some were also managed for rabbits.

The boundary between arable and pasture showed a state of flux, mainly according to short-term fluctuations in climate and to shifts in economic circumstances. Land passed into and out of cultivation according to factors such as crop failure and change in wool or cereal demand. Some of the ancient pastures have a continuous history of grazing which extends back for at least several centuries, but others went through one or more phases of ploughing and cropping, as revealed by the persistence of such features as ridge and furrow systems. An associated agricultural factor, though one related also to land tenure, was that of enclosure. Through the Middle Ages, farming was practised mainly on open-field systems, with common grazing of 'waste' land. The calcareous grasslands, in particular, were managed as open downlands and large tracts were devoid of hedges and fences. Increasingly, though, arable crops were protected against domestic stock and individual ownership units were delineated, so that field systems with boundary hedges became characteristic of the farmed lands. Open field systems declined greatly as the eighteenth century advanced, and the open common grazings of areas such as the Cotswolds and the Yorkshire Wolds became much reduced and fragmented. The Parliamentary Enclosures of 1780 to 1820, and the boost to cereal growing from the Corn Laws, further enhanced this trend and by 1850 the largest expanses of open calcareous grassland were the chalk downs of southern England from Dorset to Kent.

Up to A.D. 1600, the main expanses of lowland swamp had proved intractable to 'reclamation', though many smaller wetlands had been drained away. The East Anglian Fenland lying south of the Wash was the largest of these. When the great drainage scheme engineered by Vermuyden began in 1637, there were about 3380 km^2 of fen. By 1825 about 2500 km^2 (71 %) had been drained for agricultural use. Drainage technology became increasingly efficient, and by 1934 no more than 100 km^2 of fen remained (Thomas, Allen & Grose, 1981). Other major areas of flood plain fen, e.g. in the Somerset Levels and the valleys of the Trent and Yorkshire Ouse, were largely reclaimed to grazing marsh. Hydroseral systems with aquatic macrophytes grading to swamps of *Phragmites australis*, *Carex* spp. and carr woodland of *Alnus glutinosa* and *Salix* spp. are thus reduced to scattered remnants. The Norfolk Broads contains the largest remaining areas, but even here large expanses of fen were converted to grazing marsh.

The ancient habitats which survived most extensively in the lowlands were on acid podsolic soils and raised bog or valley bog peats with pH less than 5·0. The extreme nutrient deficiency of these substrata discouraged attempts at cultivation and, although there was local improvement and reclamation, quite large areas of acidic heathland and 'peat moss' remained in some districts. It is probably no coincidence that some of the largest remaining lowland commons and uncultivated former Royal hunting chases, such as the New Forest and Ashdown Forest, are situated on these infertile lands. On the acidic sands and clays, heaths dominated by *Calluna vulgaris*, *Erica cinerea*, *Ulex europaeus* and *Pteridium aquilinum* in varying mixtures, and locally with *Ulex gallii* and *U. minor*, were maintained by light grazing and periodic fires, but were much subjected to seral colonization by *B. pubescens*, *B. pendula* and *Pinus sylvestris*. Many heaths with wet hollows and channels retained acidophilous valley mire with *Sphagnum* spp. and vascular plants typical of ombrogenous peatland.

On some flood plains, especially in Wales, northern England and Scotland, earlier fens had developed into raised mires with considerable depths (up to 10 m) of acidic, ombrogenous peat. The largest single system was that of Thorne and Hatfield Moors south of the Humber, covering 2630 ha. These peat mosses were discouraging as an agricultural substratum but they too were gradually attacked by farmers, and the shallower ones were cut away to allow working of the underlying mineral soil The peat was useful as fuel locally and some mosses were extensively cut over for this purpose. As a result, the raised bogs of the south Lancashire coastal plains largely disappeared, and most of the remainder developed sharply defined and artificial margins, lacking the 'lagg' communities of undamaged examples in Fennoscandia. Periodic fires also tended to dry their surfaces and reduce *Sphagnum* cover: *S. imbricatum* is often abundant in peat just below the surface of these raised bogs, but scarce or absent in the living vegetation, and its decline is attributed to such disturbance.

The main uncultivated wastelands were, however, on the mountains and moorlands of northern and western Britain. By A.D. 1700, the uplands of Wales, south-west and northern England, and the Southern Uplands of Scotland had lost most of their forest cover, and since the time of the Cistercian monasteries in the twelfth century they had provided grazing range for large flocks of sheep, with numbers of goats, cattle and horses locally. The large expanses of blanket bog which covered some of the more gently contoured uplands had probably been largely treeless or only sparingly wooded during the Sub-Atlantic Period, but they were also somewhat inhospitable as grazing land.

The plagioclimax resulting from upland forest destruction evidently varied according to soil conditions and subsequent land use, and these two factors interacted. Where grazing was heavy, grasslands prevailed with species such as *Festuca ovina*, *Agrostis canina*, *A. capillaris* and *Deschampsia flexuosa* dominant on drier, skeletal brown earths, and *Nardus stricta*, *Juncus squarrosus* and *Molinia caerulea* on wetter podsols and gleys. Where grazing was lighter, there was mainly dwarf shrub heath with *Calluna vulgaris*, *Erica cinerea*, *Vaccinium myrtillus* and *V. vitis-idaea*, with local abundance of *Ulex gallii* and *Juniperus communis* on the same types of drier soil; and increasing amounts of *Erica tetralix*, *Eriophorum vaginatum*, *Eriophorum angustifolium* and *Trichophorum cespitosum* with increasing soil moisture. Soils derived from limestone and other highly base-rich rocks were preferentially grazed by the herbivores for their nutritious vegetation and developed close-cropped northern equivalents to the calcareous grasslands of the southern downlands.

It is now well known that these anthropogenic sub-montane dwarf shrub heaths have been extensively converted to the series of acidic grasslands, often with massive accompanying invasion of *Pteridium aquilinum*. Mountain areas given over largely to sheep farming, such as Snowdonia and Lakeland, became predominantly green and grassy, but with extensive bracken beds on the dry lower slopes. From around A.D. 1840 there was a new interest in managing the uplands in some areas for game shooting. On the gentler and drier moors, especially of eastern hill districts, *Calluna* was managed by rotational burning to encourage red grouse *Lagopus scoticus*, and maintained as the dominant by keeping sheep numbers low. In more rugged mountain country in the Scottish Highlands, large areas were given over to red deer *Cervus elaphus* but, since carrying capacity was generally low and the numbers of deer limited by winter food supply, the effects of grazing were less pronounced than on many sheep-walks and a good deal of dwarf shrub heath remained.

The regimes of land use outlined above continued through the nineteenth century. Reclamation of waste land to cultivation continued along previous lines, but agriculture practised a low input–low output system, and many of the man-made plant communities of farmland were floristically rich. The hay-meadow grasslands were often full of herbaceous dicotyledons which made a colourful show in flowering time, and the arable lands and farm environs had a distinctive set of weeds which were firmly established as part of our flora. World War I saw more ploughing of grassland and waste ground, but methods were much the same. The following period 1920 to 1940 was regarded as an agricultural doldrums, but significant advances in technique were being pioneered. Work at Rothamsted, Aberystwyth, Newcastle and elsewhere on plant nutrition and fertilizers, breeding of improved crop species' varieties, effects of grazing animals on sward composition, and pest control paved the way for the 'Green Revolution' in Britain.

1940 to 1984

The outbreak of World War II created an immediate need to increase domestic food production. Farming methods became modernized and Cultivation Orders led to extensive ploughing of the long-established grasslands, especially on the southern chalk downs. After the war ended, the Agriculture Act 1947 confirmed Government policy in supporting agricultural development and the goal of maximizing home food production. This set in train a renewed attrition of the remaining semi-natural vegetation of Britain, with the following effects on the main formations.

Calcareous grasslands. The large area of southern downland on chalk and Jurassic limestone was extensively ploughed and converted to arable, especially for growing barley to feed to livestock, or to commercial grass leys. Then in 1954 myxomatosis spread through Britain and reduced to very low numbers the rabbits which had also been an important grazing animal on the calcareous grasslands. Many of the remaining unploughed areas were too small for sheep grazing still to be economic, and natural seral changes occurred in the absence of grazing. First, coarse and vigorous grasses such as *Brachypodium pinnatum* and *Bromus erectus* began to luxuriate and suppress the characteristic small dicotyledonous herbs. Then the medium to tall shrubs typical of calcareous soils invaded to form rapidly thickening scrub with still greater suppressive effect on the sward plants: they included *Crataegus monogyna, Ligustrum vulgare, Cornus sanguinea, Euonymus europaeus, Rhamnus catharticus, Viburnum lantana, Rosa* spp. and *Clematis vitalba*. This is a successional stage in the development of mixed broadleaved woodland.

The outcome was that surviving calcareous grasslands tended to be on downland scarps and slopes too steep to plough, and that many of these remnants showed progressive deterioration in the absence of management. On some areas, nutrient levels were elevated by grazing cattle from richer pastures or by adding fertilizers, which can be spread by aircraft on to the steepest slopes. Crawler tractors now allow the ploughing of quite steep slopes, too, so that even these last relics have begun to dwindle.

Southern calcareous grassland is thus one of the most depleted types of natural vegetation. Few reliable survey data are available, but Jones (1973) estimated that between 1934 and 1972 chalk grassland in Dorset had declined from 7714 to 2268 ha, a reduction of 71%. Allowing that many areas not actually under cultivation have shown substantial change in floristic composition, it seems likely that no more than 20% of the total area of chalk grassland existing in 1939 survives with its former floristic richness. Many of the characteristic vascular plants have declined in parallel, e.g. *Pulsatilla vulgaris, Thesium humifusum, Blackstonia perfoliata, Iberis amara, Linum anglicum, Gentianella anglica, Senecio intergrifolius, Asperula cynanchica, Hippocrepis comosa, Phyteuma tenerum, Ajuga chamaepitys, Hypochaeris maculata, Spiranthes spiralis, Herminium monorchis, Ophrys apifera, O. insectifera,*

O. sphegodes, O. fuciflora, Himantoglossum hircinum, Anacamptis pyramidalis, Orchis simia, O. ustulata and *Aceras anthropophorum*.

Neutral grasslands. The range of grassland types within this broad category have all diminished greatly in area since 1940. Some areas have been converted to arable, but the majority have been improved, by ploughing and re-seeding with high-yielding strains of *Lolium perenne*; and on poorer soils fertilizers were also added. On the damper grasslands of marginal hill land the attempt at improvement often resulted in a great increase in the unpalatable *Juncus effusus*, and weeds such as *Cirsium arvense, C. palustre, Ranunculus acris* and *R. repens* frequently remained abundant. The development of synthetic herbicides after 1945 gave the means of producing almost weed-free grasslands, and there has been a dramatic decline in the number of hay meadows containing an abundance of grass species, *Carex* species and dicotyledon herbs. These are now reduced to scattered fields, usually surviving through the circumstance that their owners are elderly and perhaps not concerned to apply modern management techniques. The general disappearance of these old-style hay meadows has caused a great decline in species such as *Fritillaria meleagris, Colchicum autumnale, Coeloglossum viride, Orchis morio, Platanthera chlorantha, P. bifolia, Trifolium ochroleucon, Sanguisorba officinalis, Silaum silaus, Cirsium heterophyllum, Trollius europaeus, Geranium sylvaticum* and *Primula veris*.

The southern English water meadows have almost ceased to exist: they were established from A.D. 1700 to 1850 by irrigation from adjoining streams (usually on chalk) and have a distinctive floristic composition. Grazing marshes derived from earlier more hydrophilous vegetation have distinctive species such as *Caltha palustris, Galium palustre, Valeriana dioica, Pedicularis palustris, Cirsium dissectum, Lychnis flos-cuculi, Dactylorhiza incarnata, D. praetermissa, D. purpurella, Juncus articulatus, Carex hostiana, C. panicea, C. pulicaris* and *C. nigra*. These plants are still mostly widespread, but the damp grassland communities that they characterize are decreasing steadily. Surveys of sample districts have shown the following remaining proportions of unimproved pasture within total permanent grassland areas: Cumbria – 3·0%, Yorkshire Dales National Park – 4·9%, Herefordshire and Worcestershire – 5·7%, Huntingdon and Peterborough (old county) – 3·0 to 4·0% (Nature Conservancy Council, unpublished). Within the unimproved grassland category the herb-rich hay meadow is probably now the least common type.

Lowland heaths. The conversion of acidic heathland to farmland by ploughing and soil nutrient enrichment is a long-continued process. Moore (1962) found that the Dorset heaths decreased in extent from 30000 ha in 1811 to 10000 in 1960 (67% loss) mainly through this effect, afforestation and building. The heavy use of fertilizers converts infertile heathland soils to quite productive land, and reclamation has taken place widely. Farrell (unpublished) has found that for six of the main heathland areas (Dorset, Hampshire, Surrey, the Lizard, the Suffolk Sandlings and the Breckland) the overall loss between 1950 and 1984 has been 40%. Many remaining heaths not totally lost by change in land use have deteriorated appreciably through repeated fires, tree colonisation, military use and recreational pressure.

Characteristic heathland species such as *Genista anglica* and *Lycopodium clavatum* have declined markedly, but the dry heath has a rather limited flora and the more significant decrease has been in more local plants of wet heath and associated valley bog, such as *Gentiana pneumonanthe*, *Pilularia globulifera*, *Lycopodiella inundata*, *Drosera intermedia*, *Cicendia filiformis*, *Hypericum elodes*, *Scutellaria minor* and *Rhynchospora alba*. The distinctive flora of the Breckland heaths, which range from calcareous grassland to acidic heath, is also much depleted. *Holosteum umbellatum* is extinct and the characteristic disturbed ground species such as *Veronica verna*, *V. praecox*, *V. triphyllos*, *Herniaria glabra*, *Scleranthus perennis*, *Artemisia campestris* and *Crassula tillaea* are much reduced in population size.

Hedges, hedgerow trees and woods. Since 1940, one of the most noticeable changes has been in the widespread removal of hedges and their associated trees. This has mostly been done to enlarge field units so that the combine harvester can operate with maximum efficiency. The amount of hedge removal has shown wide geographical variations: it has been greatest in the cereal districts of eastern England and least in the stock-rearing country to the west and north. Pollard, Hooper & Moore (1974) have estimated that, of *c.* 500000 miles (804700 km) of hedge in Britain in 1946 to 1947, some 140000 miles (225270 km) (28%) had been removed by 1974. Hedges and their associated verges are often floristically rich and, on base-rich soils the number of shrub species forming the hedge tends to increase with age. Some hedges marking parish boundaries are extremely old and of considerable interest to historical ecologists.

The post-1940 growth of agriculture has also seen a widespread grubbing out of remaining small woods, or parts of larger woods, in order to create more farmland. This usually causes total destruction of the woodland flora on the affected ground. Kirby & Peterken (unpublished) have found that in 23 counties of England and Wales, 8878 ha out of 142228 ha (6·2%) ancient semi-natural woodland were thus destroyed during the last 50 years.

Lowland wetlands. The reclamation of lowland flood plain fens, valley mires, raised bogs and wet grazing marshes has continued during the last 45 years, though no statistics are available. Important eutrophic fens in North Wales have been reduced in area (e.g. Gors Erddreiniog on Anglesey and Gors Geirch in Gwynedd) or virtually destroyed (Gors Bodwrog on Anglesey and Ystymllyn in Gwynedd). Many remaining fens have simply dried out and progressed mainly to carr woodland through general lowering of local water tables, e.g. Newton Reigny Moss in Cumbria, Askham Bog near York and Newham Fen in Northumberland. Former fen areas such as the Somerset Levels, the Yorkshire Ouse flood meadows, Romney Marsh in Kent and the Pevensey Levels and Amberley Wild Brooks in Sussex are now mainly grazing marsh (or even arable) with aquatic vegetation only in the intersecting ditches.

Drainage operations in the interests of agriculture have also had marked effects on lowland streams and rivers. Many have been straightened and deepened, and regular cleaning out of aquatic vegetation and re-grading of banks is usual. Aquatic herbicides have also been widely used to control macrophytes. Increased run-off and water table depression have caused some lesser streams and ditches to dry out or become seasonal. Another effect of farming has been to enhance nutrient levels locally in aquatic ecosystems, through fertilizer run-off. This often leads to dominance of small green algae in adjoining lakes, ponds and ditches, and virtual disappearance of macrophytes. Some of the few basin mires remaining in Shropshire, Cheshire and Cumbria have shown extensive replacement of *Sphagnum magellanicum* by *S. recurvum*, evidently through enrichment by water from farmland within the catchments. Large numbers of ponds have been filled in on farmland as they are no longer needed for watering animals.

Uplands. Replacement of *Calluna* and *Vaccinium* heath by acidic grassland and bracken has continued in many districts. In Wales, north-west England and the Southern Uplands, progressive deterioration of heather moorland through these heavy grazing effects, combined with repeated burning, has caused the abandonment of many grouse-moors. Erosion to form scree on steep slopes, or denudation of blanket bog peat continues to extend locally, as the result of such disturbance; and indiscriminate fires cause a general drying of blanket bogs, with loss of *Sphagnum* cover, and loss of floristic variety in dwarf shrub heaths. Many northern Scottish areas of virtually natural montane heath with *Arctostaphylos alpinus*, *A. uva-ursi*, *Juniperus communis* spp. *nana* and *Loiseleuria procumbens* have been spoiled by uncontrolled moor fires during the recent run of dry summers.

The reclamation of moorland edges to provide more enclosed hill pasture or even arable has also continued in many districts. On Exmoor 23 600 ha of moorland and heath in 1947 had declined to 18 000 ha in 1976, a loss of 20% mainly to agricultural improvement (MAFF data). Between 1950 and 1980, the following areas were reclaimed in four National Parks: Dartmoor 4614 ha, Brecon Beacons 3822 ha, Northern Snowdonia 1143 ha, North York Moors 5480 ha, giving a total of 15 060 ha which it is believed can be reliably extrapolated to other regions of England and Wales to give a total of 150 000 ha reclaimed during this period (Parry, Bruce & Harkness, 1981). Figures for Scotland are not available though moorland reclamation is known to have been widespread there.

Modern techniques including the development of bracken herbicides have helped to fuel this new enthusiasm for hill land improvement, but its success depends on the repeated renewal of the fertilizer applications so necessary to boost nutrient levels in the essentially poor soils. If enough nutrients are supplied, plant growth becomes virtually hydroponic, and some of the more extreme examples of such improvement are the growing of barley on blanket bog in Sutherland and the development of grass ley on serpentine debris in Shetland.

All the above examples of expansion of farming into marginal lands rest heavily on current UK and EEC policy of open-ended public subsidy for agricultural production – a policy which has now run into serious economic difficulties and is showing signs of change accordingly.

THE DEVELOPMENT OF FORESTRY

In England, Wales and southern Scotland, most of the forest clearance had evidently taken place by A.D. 1700, and the ancient woodlands were reduced to mainly small and scattered remnants. Only a few large areas, such as the New Forest, survived for special reasons of land tenure and infertile soil. In the Scottish Highlands, deforestation reached its height during the eighteenth century, and by 1850 this too was a region of only patchy woodland cover. The remnants of the Caledonian Scots pine (*Pinus sylvestris*) forests are, nevertheless, some of our largest and finest surviving ancient woodlands. In most districts some woodland was deliberately retained as a source of firewood and material for building, fencing, tool handles, furniture and other household goods; but also as shelter and foraging ground for domestic stock. Broadleaved woodland was generally managed by coppicing, whereby the regenerating stems from cut stools were cropped on a rotation. This practice, which had been sustained in many woods over several centuries, was varied locally by allowing some of the dominant trees, especially oak *Quercus* spp., to grow into large 'standards', which were felled on a much longer rotation and used for major construction work, including shipbuilding.

The eighteenth century also saw the increase in large private country estates and, with them, the development of a new concern for woodland management, aimed partly at maintaining timber production, but also with an eye to visual and ornamental qualities. Some tree species had been introduced to Britain long before, e.g. sweet chestnut (*Castanea sativa*) but with the increasing sophistication of evolving forestry practice, numerous others were introduced, and the native species were also widely planted in areas outside their existing range. Plantation woodlands were established increasingly in place of the original native types, but there were also all variations in mixing of introduced and native trees. Sometimes the same species was replanted but with seed from another population far distant. There was also increasing selection for trees with desirable form and timber characteristics.

The overall result was that, by the end of the nineteenth century, British woodlands were not only a poor remnant of the forests that existed at the beginning of the Sub-Atlantic Period; they were also a highly confusing medley of types in ecological terms, reflecting a degree of human intervention which varied from negligible to total. Trees were also widely retained or planted singly, in lines, clumps, avenues – in all possible configurations and densities, and in both rural and urban settings. Park woodland with scattered old trees in grassland became a characteristic feature of the large country estates. Introduced species became widespread in the man-moulded environment: English elm (*Ulmus procera*), horse chestnut (*Aesculus hippocastanum*), sweet chestnut, large-leaved lime (*Tilia platyphyllos*), sycamore (*Acer pseudoplatanus*), Turkey oak (*Quercus cerris*), poplars (*Populus* spp.) and a wide variety of conifers. Some even became important woodland trees. Beech (*Fagus sylvatica*) was widely established as woodland well to the north of its existing range and Scots pine (*Pinus sylvestris*) well to the south. A very large number of exotic species were established for ornamental purposes, and commercial plantation forestry also turned increasingly to introduced species.

World War I led to the felling of some of the best remaining woods, and to the decision to create a strategic reserve of timber for the future. The Forestry Commission was established in 1919 with a brief for a massive programme of re-afforestation. The Commission's woodland census of 1924 showed that woodland then covered only 5·4% of Britain, compared with a probable area of at least 80% during the post-glacial thermal maximum. The re-planting programme sought to avoid competition with agriculture and the land acquired for afforestation tended to have predominantly infertile, acidic and base-deficient soils which farming had left as unsuitable. Some of the earliest plantations were on the acidic podsolized sands and raw chalky soils of the East Anglian Breckland, but the Commission turned increasingly to the deforested marginal lands, moorlands and hills of western and northern Britain.

Under the generally cool and humid climate of these upland districts, and through a prevalence of hard, non-calcareous parent rocks, soil conditions tended to be adverse. Podsols and various kinds of gley soils, grading into blanket bog peat of various depths, predominated, and restricted the choice of tree species for planting. Conifers were preferred, since they were fast-growing, and some species thrived on poor soils. Scots pine and Norway spruce (*Picea abies*) were the favourites at first, with larch (*Larix decidua* and *L.* × *eurolepis*) on better brown earths, and Corsican pine (*Pinus nigra* ssp. *maritima*) especially in the Breckland. It was soon found that Sitka spruce (*Picea sitchensis*) from oceanic north-west America flourished under the cool, humid climate of the British uplands, and this rapidly became the forester's favourite tree. Today, over 95% of the trees planted in Britain are of this species. When attention later turned to planting the deep peat of raised and blanket bogs, Lodgepole pine (*Pinus contorta*) was found to be the most suitable species.

Forestry has increasingly employed the methods of modern agricultural technology. Research has identified the nutritional, mycorrhizal and other ecological needs for successful establishment and growth, and plant-breeding has increasingly supplied strains with desirable performance characteristics. The ground to be planted is ploughed, nursery-grown seedlings are transplanted into the upturned sod and fertilizer (mainly ground rock phosphate) is added. Synthetic herbicides may be used to suppress vigorous competing vegetation, and insecticides to control any serious insect pest outbreaks. Fertilizer applications may be renewed by aerial spreading at intervals. The result has been to produce plantations which are the equivalent of farmland monoculture crops – large blocks of uniform age, structure and often single species composition.

During the first years after planting, removal of sheep and other stock allows the existing herbaceous and/or dwarf shrub vegetation to grow tall and flower freely, but after about 10 years the canopy closes and these communities are rapidly shaded out. They persist only in the systems of rides and fire-breaks which intersect the developing forest. The closed forest is typically dense, dark woodland almost devoid of other plant life. According to the later degree of thinning, it will later develop a field layer, but this also tends to be uniform and composed of common species such as *Pteridium aquilinum*, *Dryopteris dilatata*, *Rubus fruticosus* agg. and *Agrostis* spp. There may be a well-developed ground layer of widespread mosses. The Breckland forests have some of the best-developed and most varied field communities in modern plantation forests, and their sandy and chalky rides also support a varied flora. Under the windy, oceanic climate of the north and west, however, the increasing problem of wind-throw with advancing age of the forest is leading increasingly to a management regime which avoids thinning and even cutting of lower branches. Floristic variety is thus minimized, though new species may invade the disturbed ground of roadsides and rides, and some forests retain open ground which is too wet, rocky or elevated to plant.

The new forests thus have an artificial character which contrasts with that of semi-natural woodland still persisting under similar conditions on adjoining ground. They cannot, except in the broadest sense, be said to be a replacement of the forests which our forebears destroyed long ago. The virtues of fast-growing introduced conifers have, moreover, greatly influenced silviculture in existing woodlands of the ancient, semi-natural, broadleaved type. All over the lowlands, there has since 1940 been a widespread replacement of native species by alien conifers, with consequent reduction in structural and floristic variety. Peterken & Kirby (unpublished) found that in 23 counties of England and Wales there has been an overall loss of 40% of ancient semi-natural woodland to conifers during the last 50 years. The lowland conifer plantations are usually allowed to grow to greater maturity and are better thinned than the upland ones, but they tend to develop acidophilous communities over the needle litter, and a shrub layer is again usually absent. Sometimes, broadleaved and conifer mixtures are planted, and the conifers later removed after acting as 'nurse' to the other species. But if, through this regime, the wood passes through a thicket stage there may be floristic impoverishment similar to that under total coniferous cover. The deciduous *Nothofagus* spp. have gained popularity as commercial lowland plantation trees, but they produce dense shade inimical to field and shrub layer species.

Even when new plantations are managed more favourably for the associated communities, they do not necessarily acquire enhanced floristic variety. Many of the most characteristic woodland shrubs, herbs and ferns have very limited capacity for spread, and in fact be used as indicator species of ancient (primary) woodland (Peterken, 1974). Even widespread and locally common species such as *Mercurialis perennis* fail to cross quite short gaps in woodland cover (Peterken & Game, 1981). Woodland which has either regenerated or been planted after a significant intervening period of clearance (secondary woodland) thus tends to be floristically less varied than primary woodland. Even in broadleaved woods not subject to conifer planting, the general abandonment of coppicing for lack of economic incentive, or even of any kind of management, has led to a deterioration in ecological and floristic variety. Well-thinned woods tend to grow up into high forest, and rides close in: whilst natural regeneration after clear-felling tends to produce scrub thickets with a variety of woody species but too dense shade for many herbs. The maintenance of all seral stages between clearance and re-establishment of mature woodland is necessary if original floristic variety is to be retained whilst structural diversity and balance is highly dependent on silvicultural regime. A different problem in upland areas is that so many woods are heavily grazed and lack regeneration so that they are effectively moribund.

The overall result is that, although forest cover has now increased to 9·4% of Britain (Forestry Commision data), floristic and structural variety of our woodlands have decreased under forestry practices which increasingly amount to 'tree-farming'. National forestry policy for both public and private sectors has ensured that the increase in area since 1919 has also been almost entirely non-indigenous, coniferous woodland. Peterken (1981) estimates that ancient, semi-natural broad-leaved woodland now covers only *c.* 300000 ha, and there are another *c.* 20000 ha of native Scottish pinewood. Programmes of planting of broadleaved trees for amenity have been launched in recent years. They may eventually help to counteract the devastating ravages of Dutch elm disease, which has already killed a large proportion of the English elms in Britain and Ireland and is spreading rapidly through wych elm (*Ulmus glabra*) populations in most districts. Trees and woodland are a key part of landscape and, as the climax life-forms, they illustrate perhaps more than any other vegetational feature the dramatic impact of human activity on the face of this country.

URBAN INDUSTRIAL SPREAD

The enormous growth in human population since the Industrial Revolution, beginning *c.* A.D. 1800 has created urban sprawl and a massive system of roads, railways and, more recently, airfields. These have obviously had an obliterative effect, burying more and more land under the built environment and destroying both natural and semi-natural vegetation and farmland alike. It is estimated that *c.* 20000 ha of farmland are lost annually to such development. Green Belt policies have had only partial success in slowing the trend for conurbations to expand outwards.

The effects of such urbanization are, however, by no means wholly destructive botanically. Urban–industrial dereliction has created a great deal of waste land, colonized by a rich flora of both native and adventive species. And derelict land is still increasing. City streets yield a long list of vascular plant species, and urban parks have often retained woods of native trees with their associated flora. Golf courses, especially those in more rural settings, sometimes have a quite interesting flora. Little-modified habitats have survived in surprisingly developed areas, such as the nationally important eutrophic fen at Crymlyn Bog in industrial Swansea, Moseley Bog in Birmingham and Possil Marsh on the edge of Glasgow. As agriculture has intensified in the lowlands and destroyed ever more of the interesting herb-rich grasslands, the verges of roads and railways have become increasingly valuable as relics of such communities, with a remarkably rich flora on the more calcareous soils. Railway banks have locally helped to retain fen and marsh in adjoining farmland by ponding up water and preventing under-draining.

Reservoirs and canals have sometimes become valuable wetland habitats, the latter especially when abandoned and allowed to undergo natural plant succession. The Prees branch of the disused Shropshire Union Canal is now one of the best sites in England for aquatic macrophytes of eutrophic water.

The mineral workings associated with urban and industrial activities form an important range of botanical habitats when they are abandoned. Old quarries and mines and their spoil often develop seral stages in succession to woodland, and those cut into chalk and limestone sometimes show fine examples of herb-rich grassland with a rich calcicole flora (Davis, 1979). Distinctive old quarry species include *Ophrys apifera*, *O. insectifera*, *Anacamptis pyramidalis*, *Campanula glomerata*, *Cirsium eriophorum*, *Gentianella germanica*, *Centaurea calcitrapa*, *Ajuga chamaepitys*, *Teucrium botrys* and *Vulpia unilateralis*.

The medieval Jurassic limestone quarries at Barnack Hills and Holes, Northamptonshire were a system of shallow pits and spoil heaps which, when abandoned, became colonized by the plants from the residual matrix of grassland. The topographic irregularities prevented the agricultural improvement which destroyed the surrounding calcareous grassland, so that the area retains the only important example of this community in the whole district. Far more ancient earthworks on chalk and limestone have also preserved island fragments of calcareous grassland in an agricultural sea, in many parts of southern England. One of the best known examples is the Devil's Dyke near Newmarket, Cambridgeshire, a ribbon of chalk grassland with relict populations of *Pulsatilla vulgaris*, *Himantoglossum hircinum*, *Thesium humifusum*, *Hypochaeris maculata* and *Geranium sanguineum*. Lead mine spoil on the Carboniferous limestone in northern England has a distinctive flora with *Minuartia verna*, *Thlaspi alpestre*, *Cochlearia alpina*, *Viola lutea* and, occasionally, *Armeria maritima*. Quarries and their waste are often good habitats for ferns, bryophytes and lichens, and the two rare ferns *Asplenium septentrionale* and *A. billotii* are both locally associated with old mines cut into acidic rocks.

Some of the most valuable man-made habitats are the many large depressions produced by the working of aggregates and later abandoned. These hollows tend to fill with water and so become small lakes, often in areas with no natural standing water bodies. Mining subsidence has also produced hollows that fill with water. Plant colonization and hydroseral development usually then follow quite rapidly, the dispersal of some of the species being evidently aided by visiting waterfowl. In some disused sand and gravel pits the process has also been deliberately promoted by human introduction of appropriate plants in order to accelerate and diversify the ecological succession. Many aquatic macrophytes and fringing reedswamp species have thus been given opportunities for increase, and the creation of these artificial lakes offsets to some extent the deterioration and loss of the natural ones.

The Norfolk Broads themselves are often cited as examples of man-made wetlands, created by the medieval digging of the extensive peat deposits along the valleys of the Broadland rivers (Lambert et al., 1961). These excavations were, however, made slowly and manually at a time when the open water and fen ecosystem was extensive and otherwise primeval. There was every opportunity for colonization of the abandoned diggings by all the species concerned, which existed in abundance in immediately adjacent habitat. The essential ecological continuity necessary to spread, the close proximity of large parent populations, and the facilitation of spread by an aquatic environment, all ensured that the man-made Broads developed a vegetation of essentially natural character. This was later reinforced by the fortunate circumstance of a fall in the level of land in relation to that of the sea. Ombrogenous bogs extensively cut for fuel peat have also shown a smaller-scale regeneration of actively growing surface, a good example being the basin mire of Moorthwaite Moss, Cumbria, which redeveloped a *Sphagnum*-dominated hummock and hollow system in old cuttings.

Walls and buildings have become an important botanical habitat, though older examples and those in villages and rural settings have most interest. Drystone walls have a diverse bryophyte and lichen flora, varying from calcifuge to calcicole according to lime content of the parent material. Mortar supplies $CaCO_3$ to walls made of acidic materials, and produces a characteristic fern flora with *Ceterach officinarum*, *Asplenium trichomanes*, *A. ruta-muraria*, *A. adiantum-nigrum*, *Phyllitis scolopendrium* and *Polypodium cambricum*: the first three species are more abundant and luxuriant on walls than in any natural habitat in Britain. Several species of both native and introduced vascular species are highly characteristic of walls, e.g. *Parietaria judaica*, *Geranium lucidum*, *G. robertianum*, *Cymbalaria muralis*, *Centranthus ruber*, *Erinus alpinus*, *Cheiranthus cheiri*, *Mycelis muralis*, *Chelidonium majus*, *Corydalis lutea*, *Sonchus oleraceus*, *Sedum* spp., *Hieracium* spp. and *Umbilicus rupestris* (acidic rocks). Distinctive wall communities have been recognized (Darlington, 1981) and there is a characteristic bryophyte and lichen flora of walls, in which calcifuge and calcicole elements are clearly differentiated according to rock type and presence or absence of mortar.

Probably the more serious and intractable effect of urban–industrial development is the less direct influence of atmospheric pollution from large-scale combustion of fossil fuels. The Industrial Revolution, founded on coal instead of wood as a fuel, produced massive amounts of soot and acidic solid particles. The gaseous sulphur dioxide and nitrogen oxides became converted in rain and deposited as dilute sulphuric and nitric acids. Oil, which much later (mainly post-1950), became another major fuel, produces less soot but on average has similar acidifying effects as coal. Conway (1949) connected the virtual disappearance of the once-abundant peat-forming Sphagna from the blanket bogs of the Southern Pennines with the heavy fall-out of acidic emissions from the large industrial complexes to west and east. Some ecologists also attributed the extensive erosion of these bogs, and the usual dominance of *Vaccinium myrtillus* and *Empetrum nigrum* to the same cause.

Lichens have also long been known as plants particularly sensitive to acidity induced by atmospheric pollution. Hawksworth, Rose & Coppins (1973) have shown the clear evidence for progressive impoverishment of lichen floras with proximity to urban–industrial pollution sources. Many species have undoubtedly declined markedly, or even become extinct, in the most heavily populated and industrialized parts of Britain (Seaward & Hitch, 1982). The effect explains why the Lake District, immediately in the lee of the Workington and Barrow-in-Furness industrial areas, is now less rich in lichens than Snowdonia, an otherwise similar district ecologically. A wide range of species' tolerance has, however, been revealed within the concentric patterns of lichen distribution around major pollution sources, and the distribution limits of many species correlate closely with particular levels of sulphur dioxide air pollution. Some of the most sensitive species are the large foliose lichens of the distinctive community *Lobarion pulmonariae*, which occurs especially on the bark of old trees but also on rocks in western districts: they include all species of *Lobaria*, *Sticta*, *Pannaria* and *Parmeliella*. *Parmelia caperata* and *Evernia prunastri* are examples of species with intermediate tolerance of acidity, which have declined only in the most polluted districts. *Lecanora conizaeoides* has a high acidity tolerance and has actually spread markedly in Britain since 1960 (Seaward & Hitch, 1982).

Although they have been less closely studied, there are clear indications that some species of mosses and liverworts which grow on trees have shown the same pattern of decline as lichens. They include most species in the mainly arboreal genera *Ulota* and *Orthotrichum*, *Leucodon sciuroides*, *Tortula laevipila*, *Frullania tamarisci* and *F. dilatata*.

While there is a lower output of atmospheric pollution generally, and sulphur dioxide emissions have declined since 1955, the problem of acid deposition is believed to have become worse. It is, indeed, one of the most topical conservation issues in Europe and North America. Most industrial processes and many urban users now rely on electricity for energy, and this has localized the main sources of acid emission to the huge power stations which have multiplied since 1950. Construction involved a policy of building tall smoke stacks whose effect is to reduce local acid deposition but to spread this over a larger area to leeward. There is much current debate over assertions that acid deposition, mainly from industrial sources in Britain and continental Europe, is causing widespread death and damage to coniferous forests in central Europe and to plants and animal life in lakes in southern Fennoscandia.

In Britain, argument has been partly over whether the long-established and well-known levels and effects of acid deposition following the Industrial Revolution have shown any worsening in recent years. Flower & Battarbee (1983) have shown that diatom composition in cores from certain Galloway lakes gives evidence of further post-1940 enhancement in the trend of acidification which began around 1840. It is generally accepted that acidification is most marked in freshwater ecosystems, especially lakes, in upland areas of predominantly acidic parent rocks and soils where buffering capacity is lowest. There is also an increase of acidification in extensive areas of coniferous afforestation, resulting partly from the ability of the foliage to trap aerosol particles. A recent review of the available evidence (Fry & Cooke, 1984) has shown that the main effects which pose conservation problems are the possible spread of acidification into districts of western Scotland containing internationally important hyper-oceanic lichen communities, and the loss in variety of animals and perhaps plant species in some of the already acidic lakes and rivers of some western districts.

Another serious problem in urban–industrial pollution is the discharge of waste into rivers, causing either direct toxic effects or eutrophication through nutrient enrichment. There have been considerable and relatively successful efforts to clean up river pollution during recent years and beneficial effects on fish populations have been noted. There has, however, been a widespread loss of aquatic macrophytes in both rivers and lakes, associated with blooms of green algae, attributable to the run-off of agricultural fertilizers, deliberate herbicide treatment in the interests of land drainage and river navigation, and sewage discharge. The Norfolk Broads are perhaps the best-known case, with dieback of the extensive beds of reed *Phragmites australis* an additional problem. Careful studies here have shown an extremely complex situation, with contributory factors including discharge of sewage from domestic sources and holiday boats, fertilizer run-off, enrichment from a large black-headed gull (*Larus ridibundus*) roost, disturbance of banks and bottom muds by boats, seepage of saline water under tidal influence and iron ochre release from pyritic peats (Moss, 1983). Different factors may exert a predominant influence in different localities.

Other important lakes showing evidence of eutrophication are Llangorse Lake and Bosherston Lake in South Wales, Malham Tarn in Yorkshire and Loch Leven in Scotland. While these probably involve certain influences common to all instances of eutrophication, remedial action involves tracing precisely the sources of enrichment, so that each case requires a separate study.

Oil pollution at sea has damaging effects on the intertidal communities of the coastal habitats where drifted oil washes ashore – the rocky littoral, sediment flats, shingle beaches and salt marshes. Use of oil dispersants can cause further damage. Natural recolonization usually restores the former vegetation, but may take many years on badly affected shores, recovery being slowest on shores sheltered from wave action.

The needs of industry and the city have other far distant effects. Water supply is a major environmental influence, and numerous reservoirs have been made, either by raising the level of existing lakes or damming river valleys. The characteristic zonation of vegetation in natural lakes is usually destroyed by artificial raising of water level, and the terrestrial vegetation of flooded valleys is simply drowned. Since most reservoirs have a marked draw-down zone with frequently fluctuating water levels, marginal vegetation often fails to re-establish. Rare species lost in this way are *Juncus filiformis* beside Thirlmere, *Schoenus ferrugineus* on Loch Tummel and *Saxifraga hirculus* in Baldersdale.

The most serious case was the Cow Green reservoir in Upper Teesdale. This much-contested proposal to supply water to industrial Teesmouth submerged about one-quarter of the area of unique sugar limestone grassland and flush vegetation on Widdybank Fell. Significant fractions of the populations of rare montane species such as *Viola rupestris* and *Kobresia simpliciuscula* were drowned, and acidophilous bog communities with the rare *Carex paupercula*, *C. aquatilis* and *Sphagnum riparium* were also lost. The only gain was that the reservoir promoters, Imperial Chemical Industries Ltd, set up a substantial research fund which enlarged knowledge of Upper Teesdale botany (Clapham, 1978).

Water engineering has increased greatly in sophistication and capacity. There is concern amongst freshwater ecologists over proposals for inter-catchment transfers of water, e.g. from the Severn to the Trent. Water locally supplies hydro-electric power, mainly in the mountains of Wales and Scotland. While its adverse effects are mainly on landscape, the developments on Ben Lawers damaged interesting botanical ground and provided access roads which led to increased collecting of rare alpines. And in Snowdonia, the breaching of the Llyn Eigiau dam in 1925 scoured the Porth-Llwyd chasm, one of the finest waterfall ravines for Atlantic bryophytes in Britain.

Nuclear power stations are built away from human population centres, and that on the Dungeness promotory was damaging to Europe's finest system of coastal shingle ridges, with distinctive sub-maritime plant communities. Proposals for estuarine barrages to harness tidal power have not yet materialized but would cause a substantial loss of salt marshes and intertidal flats, and their replacement by shallow freshwater or terrestrial habitats.

Defence uses of land have beneficial effects to semi-natural vegetation on the whole, for they often prevent other adverse developments, notably for agricultural purposes. By far the largest areas of chalk grassland now left are on the Ministry of Defence ranges at Porton Down and on Salisbury Plain in Wiltshire. In east Anglia, the Stanford Training Area is much the largest remaining single expanse of Breckland grass-heath. In these areas it is possible to see not only the distinctive plant communities and flora, but whole landscapes characterized by these types as they existed widely up to 1940.

Recreation

Landowners' enthusiasm for hunting and shooting was responsible for maintaining many lowland native woodlands as fox and pheasant covert, and large areas of mountain, moorland and blanket bog as deer forest and grouse moor. These interests protected the areas concerned from the more intensive land use developments stemming especially from agriculture, both lowland and upland. Angling interests have also been concerned to maintain water quality in lakes and rivers, and have to some extent restrained freshwater pollution and its damaging effects on aquatic vegetation.

The post-1945 era of increasing public access to the countryside has seen a locally pronounced impact on certain habitats such as chalk downs, acidic heaths, woodlands, waterfall glens, sand dunes, mountains and moorlands. Trampling of popular sites and especially paths has caused quite severe erosion of vegetation and soil: particularly serious examples occur on the blanket bogs crossed by the Pennine Way. Fire risks are increased in areas with large numbers of people and accidental moorland fires during dry summers are another local cause of soil and peat erosion. Holiday boating is another form of disturbance which is most marked in the Norfolk Broads, where it is regarded as a contributory factor in the recent deterioration of the hydroseral system (p. 91).

The collecting of wild plants has had quite an important effect on the British flora. Although an essential part of the Golden Age of botanical taxonomy, the compilation of herbaria increasingly became a pastime, with rare species regarded as particular prizes. Much living material was also uprooted for the garden or nursery. The outcome is that today some of our rarer species are much depleted in total populations, and some have been made locally extinct. Ferns, orchids and montane plants have suffered especially from collecting, and several species of each group are amongst the most endangered members of the British flora. The disappearance of *Spiranthes aestivalis* from its only station is attributed partly to collecting and *Cypripedium calceolus* verges on national extinction. Other orchids whose survival is threatened by collecting include *Orchis militaris*, *O. simia*, *Epipogium aphyllum* and *Cephalanthera rubra*. Ferns much depleted, at least locally, are *Osmunda regalis*, *Adiantum capillus-veneris*, *Trichomanes speciosum*,

Asplenium bilottii, *A. septentrionale*, *Cystopteris montana*, *Woodsia ilvensis*, *W. alpina* and *Polystichum lonchitis*. Reduced montane-northern species are *Lychnis alpina*, *Potentilla rupestris*, *Saxifraga cernua*, *S. cespitosa*, *Gentiana verna* and *Lloydia serotina*. Some rare bryophytes and lichens have also been reduced in abundance through over-collecting in certain well-known localities.

Many other rare and local species are over-represented in herbaria, yet are still tolerably plentiful in the named localities: either they were once more abundant still, or they have replenished their populations successfully. The other kind of collecting is of common, but colourful and attractive flowers. In some districts *Primula vulgaris* is now quite uncommon on roadside banks through uprooting, but picking of the flowers of species such as *Narcissus pseudonarcissus* may actually enccurage vegetative growth.

The enormous increase in popularity of gardening has had quite profound effects on habitat and vegetation. The dried and milled peat from ombrogenous bogs is a valued propagating medium and soil conditioner, especially with added nutrients. The resulting demand for horticultural 'moss litter' has led to large-scale commercial cutting of peat, especially that composed of *Sphagnum*. Some large raised bogs previously worked for fuel peat have now been extensively cut-over for moss litter and their vegetation greatly modified. Cutting is usually by trenches, leaving dry ridges between so that hydrophilous species such as Sphagna tend to disappear and only plants tolerant of dry conditions (e.g. *Calluna vulgaris*) persist. Regeneration may eventually occur in abandoned cuttings, but draining has sometimes lowered water tables too much, and the original surface structure of the bog has been destroyed. Important raised bogs extensively modified in this way include Shapwick Heath, Somerset; Thorne and Hatfield Moors, Humberside; and Wedholme Flow, Solway Moss and Bolton Fell, Cumbria.

Another important effect has been through the demand for the weathered surface layers of Carboniferous limestone 'pavements' as ornamental and rockery stone. These irreplaceable geological features with tabular, variably fissured surfaces are extremely localized in Britain, and occur mainly in the Craven Pennines of Yorkshire, Cumbria and North Lancashire. They are of great botanical interest, with a distinctive calcicole flora containing many rare and local species, and showing a wide ecological range of communities from woodland and scrub to open sub-montane types. Ward & Evans (1976) surveyed all the 537 major occurrences of limestone pavement in Britain, and found that 39% of their total extent had been damaged or destroyed, mainly by recent removal of surface stone. Only 3% of the 537 pavements showed no detectable damage, and only 13% were considered to be 95% intact. Stone removal on some pavements has continued, extending damage to an estimated 45%.

Gardening has, however, had the effect of adding many species to the flora growing wild in this country.

THE INTRODUCED FLORA

New species have been introduced to Britain by man over a long period, increasingly since the Middle Ages. Some were brought deliberately for functional or aesthetic purposes while others came accidentally. Many species have remained localized, e.g. around ports, while others have become widespread in suitable habitats. Because evidence of origin varies greatly, there is a virtual continuum between species known to be introductions and those probably native, at least in some localities; but with a considerably 'grey area' in the middle where argument over such status is both academic and sterile. The variety of terms which has been used for varying categories of non-native plants is thus of doubtful utility. Perring & Walters (1962) indicated 239 taxa as introductions, and another 363 as probable or certain natives whose distribution has been artificially extended.

Many introduced species now have an important place in both our flora and plant communities, and a few are vigorous competitors which replace certain natives. One of the most problematical is *Rhododendron ponticum*, so widely planted as an ornamental shrub and now rampant in dense thickets through some of our best remaining western oakwoods. It is steadily spreading through the Killarney oakwoods with their important fern, bryophyte and lichen communities. *Reynoutria japonica* is another vigorous invasive shrub lately on the advance in waste places.

Many lowland woods managed as pheasant coverts have growths of *Mahonia aquifolium*, *Symphoricarpus rivularis* and *Gaultheria shallon* planted for food and cover. Berry-bearing trees and shrubs in gardens are often spread around by birds, and so appear in quite different, even natural habitats such as cliffs and steep rocky ground, e.g. *Berberis vulgaris*, *Cotoneaster simonsii*, *C. microphylla*, *Sorbus aria* and *S. intermedia*. Woodland and scrub have various species native in some localities but introduced in many others (including a wider variety of habitats), such as *Aquilegia vulgaris*, *Convallaria majalis*, *Atropa belladonna*, *Aconitum anglicum*, *Helleborus viridis* and *H. foetidus*.

On the coast, the shrub *Hippophaë rhamnoides* is native on some dune systems but introduced on others. It becomes locally dominant in places and is generally regarded as a nuisance which should be controlled. The New Zealand herb *Acaena anserinifolia* was introduced with wool consignments and replaces native species on some sand dunes. One of the most celebrated introductions was of the North American *Spartina alterniflora* to Southampton Water, where it spontaneously hybridized with the native *S. maritima* to produce the F_1 hybrid, *S.* × *townsendii*, and subsequently the vigorous tetraploid *S. anglica*. The former was first recorded in about 1870 and the latter arose about 20 years later (Goodman et al., 1969). They proved most effective mudflat stabilizers and were introduced to various other estuaries for this purpose. They spread to others still, and have to some extent replaced the pioneer salt marsh seral stages in these areas, for their dense growths dominate species such as *Salicornia* spp., *Puccinellia maritima* and other halophytes.

Some aquatic or semi-aquatic plants have assumed an important place in our freshwater vegetation. *Elodea canadensis* was introduced to Ireland in 1836 and then to Britain in 1842. It spread rapidly in nutrient-rich waterways so abundantly as to choke many of them, but then declined mysteriously. It is still widespread but not usually abundant. Another North American aquatic, *Azolla filiculoides* is naturalized in many ponds and ditches in southern England, but becomes reduced in hard winters. *Acorus calamus* grows in many southern hydroseres, *Impatiens glandulifera* forms dense fringing growths along lowland river banks, *Lupinus nootkatensis* makes large patches on Highland river shingle and *Mimulus guttatus* thrives along smaller streamlets. In the southern mountains, and especially in Ireland, the New Zealand *Epilobium brunnescens* has abundantly colonized moist rocks and open places, and appears to compete with native species in places.

Gardens have been a fruitful source of 'escapes' of hardier species, which flourish in waste places, roadside and railway verges, and are now widespread species of such habitats. *Senecio squalidus* has made a particularly dramatic spread along railways and in towns during recent years and *Buddleia davidii* is widely established on railway waste ground and walls. Species cultivated as herbs are usually in such habitats not far from buildings, e.g. *Myrrhis odorata, Foeniculum vulgare, Rumex alpinus* and *Polygonum bistorta*.

Arable land had, over the centuries, developed a large and characteristic flora of herbaceous 'weeds', many of then annuals. Many were introductions but some were of ancient and perhaps native origin. The developing practice, during the last few decades, of cleaning the seed of crop plants for sowing, as well as the use of herbicides, has gradually reduced or eliminated these weeds, especially in cornfields. Although poppies *Papaver* spp. still produce spectacular displays here and there every year, many once common cereal weeds have declined greatly, e.g. *Agrostemma githago* (possibly heading for extinction), *Centaurea cyanus, Chrysanthemum segetum, Adonis annua, Myosurus minimus, Scandix pecten-veneris, Lithospermum arvense, Legousia hybrida* and *Bromus secalinus*. Species once widely grown as crops have persisted widely in roadside verges and other farmland habitats, but are evidently declining, e.g. *Linum usitatissimum, Cichorium intybus, Onobrychis viciifolia, Medicago sativa* and *Fagopyrum esculentum*. Weeds in general have decreased through the recent enthusiasm for the many synthetic herbicides now available.

Decline in the British Flora

Of the vascular plant species recorded from living material in Britain, only 19 are known to have become extinct. Those lost before 1930 were *Euphorbia villosa, Centaurium latifolium, Pinguicula alpina, Senecio congestus, Scirpus hudsonianus, Carex davalliana* and *Rubus arcticus*; and those lost after 1930 are *Holosteum umbellatum, Hydrilla verticillata, Halimione pedunculata, Campanula persicifolia, Saxifraga rosacea, Spiranthes aestivalis, Otanthus maritimus, Arnoseris minima, Bupleurum rotundifolium, Bromus interruptus* and the introduced *Filago gallica* and *Ajuga genevensis*. Most of these extinctions are attributable to human agency (Perring & Farrell, 1983). At least two fen mosses, *Paludella squarrosa* and *Helodium blandowii*, are regarded as extinct.

Perring and Walters (1962) showed that a large number of vascular species declined between 1930 and 1960, and many of these are known to have declined further subsequently, in parallel with their habitats, as described above. Tables 1 and 2 give some indication of the trend, which is likely to continue. New species will, predictably, be slow to arrive of their own accord, and are most likely to be those with propagules carried by wind over long distances (mainly orchids, ferns, bryophytes, lichens and fungi) or readily dispersed by birds.

The Conservation of Vegetation and Flora

The twentieth century has seen increasing concern to conserve our remaining vegetation and wild flora against further inroads by development. Progress was slow until the setting up in 1949 of the Nature Conservancy (which became the Nature Conservancy Council in 1973) as the Government's agency for the conservation of wildlife and physical features. The voluntary movement for nature conservation has since increased enormously in size and effectiveness, and now numbers at least one million supporters. The Conservancy has established a series

Table 1. *Nationally rare species of British vascular plant showing at least 33% decline since 1930**

Habitat category	Number of species	Percentage of total (317 species)
Dry banks, shingle	14	4·4
Arable	12	3·8
Woodlands	8	2·5
Grassland	34	10·7
Sandy areas/heaths	19	6·0
Upland	9	2·8
Wetland	21	6·6
Total	117	36·9

* Tables 1 and 2 have been compiled by L. Farrell, NCC, from the maps showing pre- and post-1930 distribution of British vascular plants in the *Atlas of the British Flora* (Perring & Walters, 1962). Records are at the scale of 10 × 10 km squares of the National Grid. Nationally rare species (317 in total) are those occurring in 15 or fewer grid squares and listed in Perring & Farrell (1983). Species in Table 2 are the remainder of the total of 1423 British native vascular species (excluding micro-species and known introductions).

Table 2. *Other species of British vascular plant showing at least 20% decline since 1930**

Habitat category	Number of species	Percentage of British native flora
Dry banks, shingle	3	0·3
Arable	6	0·4
Woodlands	18	1·3
Grassland	32	2·2
Sandy areas/heaths	14	1·0
Upland	7	0·5
Wetland	69	4·8
Total	149	10·5

of 195 National Nature Reserves covering c. 150000 ha and representing the countrywide range of variation in natural and semi-natural ecosystems with their associated assemblages of plant and animal species. Another 600+ areas also deserving the status of National Nature Reserves have been identified, and are included within a total of over 3000 Sites of Special Scientific Interest notified for their biological features. Many other nature reserves have been set up by other bodies, and at present about 7% of Britain is under some degree of protection for its 'nature interest'.

Increasingly, the conservation strategy has been to protect by reserve status as much as possible of remaining semi-natural habitat in the lowlands, and to ensure that in the uplands and on the coast a reasonable number of fairly large areas out of the still huge extent of undeveloped land is given adequate safeguard. Some reserves have been chosen for floristic rather than vegetational features, e.g. important assemblages of species and rare species. Populations of the majority of native vascular species are represented in at least one reserve, and many of the important plant refugia, such as Upper Teesdale and Ben Lawers, are protected. Man-made habitats ranging from flooded gravel pits to strips of roadside verge are included in the lowland series.

Nature reserves have to be managed to maintain their biological interest. Our woodlands are mostly so small that continuous interevention through such methods as coppicing is necessary to maintain all stages in the woodland succession, especially the floristically rich communities of non-shaded ground. Grasslands have evolved under a long-continued grazing regime, and this has to be maintained within quite precise limits if desired floristic composition is to be kept. The management of remaining lowland calcareous grasslands has been a special problem after myxomatosis removed rabbits and grazing by domestic stock became uneconomic. Seral development of scrub and woodland on both grasslands and heaths has to be kept in check, and carr invasion of fen reed and sedge swamp under falling water tables also has to be controlled. Invasive introduced species which suppress native communities (e.g. *Rhododendron ponticum*) may need eradicating, but this can be difficult. Freshwater ecosystems have to be maintained against eutrophication or other deterioration in water quality, and this is even more problematical when the sources of pollution lie outside the actual reserve. Protected areas also have to be wardened to prevent deliberate damage, through fires, trampling disturbance to sensitive communities and plant collecting.

Nature conservation strategy also seeks to minimize further losses of wildlife and its habitat in the wider countryside. This has to be mainly through persuading landowners and occupiers, and other natural resource managers, of the need to take due account of nature conservation in pursuance of their main interests, and to avoid unnecessary destruction and damage to semi-natural vegetation and wild plants. It tries to soften the damaging impact on nature of other Government policies on land use, and to encourage local authorities, National Park Authorities and other administering bodies to develop their own strategies for nature conservation. This wider approach relies heavily on education and publicity, directed at all levels from school children to national policy makers, and seeks especially to instil awareness and knowledge of nature, ecological processes and resource conservation issues. It promotes the view of wild plants and animals with their physical environment as part of the nation's natural heritage. Their conservation accordingly deserves to be regarded as one of the major resource uses for a spectrum of purposes: economic, scientific, educational, recreational and aesthetic.

The main nature conservation problems in Britain stem from attempts to increase the yield of material products from the land to satisfy the need for economic growth. Production of food and timber are increased by modern techniques which aim to maximize the energy and nutrient flows in the ecosystem through the crop: an intensification process which allows correspondingly less of the biomass to be produced by the ancillary species of both plant and animal. Both farming and forestry have tended increasingly to develop true monoculture systems. Other uses of natural resources associated with urban development and the built environment all tend to disrupt or displace original ecosystems. Nature conservation involves placing constraints on these forms of growth. There is thus scope for confusion with the holistic or integrated concept of natural resource conservation developed in the World Conservation Strategy (Anon, 1980). Nature conservation, in the sense described above, is one sectional element within this broader view of environmental management for all the needs of mankind. In Britain, too, the rationale of conserving genetic diversity as an insurance against the future has less force than in countries where a higher proportion of commercial plants and animals are derived from wild indigenous forms.

Nature conservation practice has to be underpinned by an information system derived from research. This begins with surveys to record the range of variation in vegetation and flora, and its distribution and abundance, as the basis for evaluation for site selection and other elements of the programme. Monitoring schemes (see p. 77) are necessary to detect and measure change in botanical features and thus to help identify problems and priorities. And ecological research is needed to understand relationships and processes, so as to give insight for decisions and for management in the field.

Vegetation classification is especially relevant to survey and mapping, and the characterization of important sites. Monitoring of taxonomic groups, (cf. Perring & Walters 1962), helps to identify declining and threatened species. These can then be given closer attention, including exact population counts, and brought to particular attention through the Red Data Book series, e.g. Perring & Farrell, 1983. Botanic gardens have co-operated in the conservation of rare species by taking material into cultivation, and the seed bank at Kew has also accepted seed of many species. Wild strains of commercially developed grasses and other crop species are also being maintained in reserves and other sites, so that their genetic potential is secured for the future.

An increasingly important aspect in future will be the re-creation of plant communities and re-introduction of species which have declined or disappeared locally. Techniques have been evolved for the re-establishment of herb-rich neutral grassland, and suitable seeds mixtures are now being produced by commercial firms. The recolonization of flooded old gravel pits has been accelerated by species introductions, and the re-vegetation of industrial waste and mineral extraction sites is taking place widely, using methods and species developed by ecologists for dealing with intractable substrata. Reserves with arable weeds have been established to represent this disappearing flora. Fen and even ombrogenous bog communities are being redeveloped in cut-over lowland bogs, and new lakes and pools created in suitable places. New afforestation offers considerable scope for this creative conservation, depending on willingness to forgo maximum possible timber production.

Nature conservation has to work within a statutory framework of available measures and restraints. The Wildlife and Countryside Act 1981 has strengthened the provisions for safeguarding of Sites of Special Scientific Importance, and given the power to establish Marine Nature Reserves, and to protect limestone pavements. It forbids the picking or uprooting of any wild plants without permission of the owner or occupier, and protects under special penalties 61 listed vascular species vulnerable to collecting and hence extinction through rarity or attractiveness. The schedule of vulnerable species has to be reviewed at 5-year intervals, and the legislation in general is monitored for effectiveness, with a view to further improvement at the next opportunity. A related input is the surveillance of socio-economic policies and trends which so much influence the many aspects of human environmental impact. And the final aim is to obtain increasing public support for nature conservation, as an integral part of the nation's concerns (Nature Conservancy Council, 1984).

CONCLUSIONS

Vegetational change during the last 300 years has been so dominated by human influence that possible effects of climatic fluctuation are barely detectable. The rapidly expanding development of land and other natural resources for human purposes has increasingly covered Britain with man-made vegetation. Natural vegetation is now reduced to scattered and insignificant fragments and even semi-natural plant communities now cover no more than 30 to 35 % of our land surface, mainly on the western and northern uplands. The built environment, with an unknown area of gardens, parks and waste ground, covers about 8 % of Britain and woodland 9·4 %, of which 70 % is alien conifer plantation. Arable croplands account for 32 % and permanent enclosed grassland 22 % of the countryside.

At present rates of expansion, the urban area will have increased to over 11 % by A.D. 2000, and the forest area to a similar amount. Agricultural land area may decrease slightly through such expansion and because economic forces are likely to militate increasingly against the recent open-ended reclamation of marginal land. Nature conservation measures aim to increase the present 7 % of natural and semi-natural habitats (i.e. vegetation) protected and managed, both for cultural (including scientific) and economic purposes, to at least 10 %. Existing land use will be compatible with maintenance of nature interest over much of this area, but development will be restricted. Other measures will aim to conserve the wild flora in the generality of the rural and urban environment, and to enhance the resources of nature through creative conservation.

The character of British vegetation and flora will, accordingly, become ever more tightly controlled by conscious human intervention and dependent on deliberate choice. The natural processes of plant succession, spread and retreat will continue but, with still further increases in manipulation, their importance will probably decline. Our aims should thus include ensuring that there remain in perpetuity areas where the process of nature, free from otherwise all-pervading human influence, can still be observed, moulding the character of our vegetation and its constituent flora.

References

Anon (1980). *World Conservation Strategy: Living resource conservation for sustainable development.* International Union for the Conservation of Nature and Natural Resources.

Aas, B. (1969). Climatically raised birch lines in south eastern Norway 1918–1968. *Norsk Geografisk Tidsskrift*, **23**, 119–130.

Clapham, A. R. (ed.) (1978). *Upper Teesdale. The Area and its Natural History.* Collins, London.

Conway, V. M. (1949). Ringinglow Bog, near Sheffield, Part II. The present surface. *Journal of Ecology*, **37**, 148–170.

Darlington, A. (1981). *Ecology of Walls.* Heinemann, London.

Davis, B. N. K. (1979). Chalk and limestone quarries as wildlife habitats. *Minerals and the Environment*, **1**, 48–56.

Erkamo, V. (1952). On plant-biological phenomena accompanying the present climatic change. *Fennia*, **75**, 25–37.

Flower, R. J. & Battarbee, R. N. (1983). Diatom evidence for recent acidification of two Scottish lochs. *Nature*, **305**, 130–132.

Ford, M. J. (1982). *The Changing Climate: Responses of the Natural Fauna and Flora.* Allen and Unwin, London.

Fry, G. L. A. & Cooke, A. S. (1984). *Acid Deposition and its Implications for Nature Conservation in Britain.* Focus on nature conservation, No. 7. Nature Conservancy Council.

Godwin, H. (1975). *The History of the British Flora.* 2nd Edn. Cambridge University Press, Cambridge.

Good, R. (1936). On the distribution of the lizard orchid (*Himantoglossum hircinum* Koch). *The New Phytologist* **35**, 142–170.

Goodman, P. J., Braybrooks, E. M., Marchant, C. J. & Lambert, J. M. (1969). Biological flora of the British Isles. *Spartina* × *townsendii* H. & J. Groves *sensu lato*. *Journal of Ecology*, **57**, 298–313.

Hawksworth, D. L., Rose, F. & Coppins, B. J. (1973). Changes in the lichen flora of England and Wales attributable to pollution of the air by sulphur dioxide. In: *Air Pollution and Lichens* (ed. by B. W. Ferry, M. S. Baddeley & D. L. Hawksworth), pp. 330–367. Athlone Press, London.

HEARN, K. (1976). The effect of the 1976 drought on sites of nature conservation interest in England and Wales. In: *Effects of the 1975–1976 Drought on Wildlife* (Ed. by D. J. Hill & R. A. Avery), pp. 45–55 University of Bristol.

JERMY, A. C., ARNOLD, H. R., FARRELL, L. & PERRING, F. H. (1978). *Atlas of Ferns of the British Isles*. The Botanical Society of the British Isles and the British Pteridological Society.

JONES, C. (1973). *The Conservation of Chalk Grassland in Dorset*. Dorset County Council.

KULLMAN, L. (1979). Change and stability in the altitude of the birch tree-limit in the southern Swedish Scandes 1915–1975. *Acta Phytogeographica Suecica* **65**, Uppsala.

KULLMAN, L. (1981). Recent tree-limit dynamics of Scots pine (*Pinus sylvestris* L.) in the southern Swedish Scandes. *Wahlenbergia*, **8**, 1–67.

LAMB, H. H. (1982). *Climate, History and the Modern World*. Methuen, London.

LAMBERT, J. M., JENNINGS, J. N., SMITH, C. T., GREEN, C. & HUTCHINSON, J. N. (1961). *The Making of the Broads. Royal Geographical Society* Research Memoir No. 3. London.

MOORE, N. W. (1962). The heaths of Dorset and their conservation. *Journal of Ecology* **50**, 369–391.

Moss, B. (1983). Norfolk Broadland; experiments in the restoration of a complex wetland. *Biological Reviews* **58**, 521–567.

NATURE CONSERVANCY COUNCIL (1984). *Nature Conservation in Great Britain*. NCC, London.

PARRY, M., BRUCE, A. & HARKNESS, C. (1981). The plight of British Moorlands. *New Scientist*, **90**, no. 1255, 550–551.

PERRING, F. H. & FARRELL, L. (1983). *British Red Data Book I. Vascular Plants*, 2nd Ed. RSNC, Lincoln.

PERRING, F. H. & WALTERS, S. M. (ed.) (1962). *Atlas of the British Flora*, Nelson, London.

PETERKEN, G. F. (1974). A method for assessing woodland flora for conservation using indicator species. *Biological Conservation* **6**, 239–245.

PETERKEN, G. F. (1981). *Woodland Conservation and Management*. Chapman and Hall, London.

PETERKEN, G. F. & GAME, M. (1981). Historical factors affecting the distribution of *Mercurialis perennis* in Central Lincolnshire. *Journal of Ecology*, **60**, 781–796.

POLLARD, E., HOOPER M. D. & MOORE, N. W. (1974). *Hedges*. New Naturalist Series, Collins, London.

SEWARD, M. R. D. & HITCH, C. J. B. (1982). *Atlas of Lichens of the British Isles*. Vol. 1. Institute of Terrestrial Ecology (NERC), Cambridge.

TANSLEY, A. G. (1939). *The British Islands and their Vegetation*. Cambridge University Press, Cambridge.

THOMAS, G. J., ALLEN, D. A. & GROSE, M. P. B. (1981). The demography and flora of the Ouse Washes, England. *Biological Conservation*, **21**, 197–229.

WARD, S. D. & EVANS, D. F. (1976). Conservation assessment of British limestone pavements based on floristic criteria. *Biological Conservation*, **9**, 217–233.

CYTOGENETIC VARIATION IN THE BRITISH FLORA: ORIGINS AND SIGNIFICANCE

By T. T. ELKINGTON

Department of Botany, The University, Sheffield, S10 2TN, UK

SUMMARY

Cytogenetic variation in the British flora is discussed in relation to the taxonomy and evolution of species and to their distribution and past history. In polyploids there may be two or more cytotypes in Britain with different distributions and ecology, e.g. *Empetrum nigrum* and *Hippocrepis comosa*; alternatively only one cytotype is present in Britain, but it forms part of a wider polyploid pattern, the study of which may clarify past patterns of migration, e.g. *Potentilla fruticosa* and *Sisyrinchium bermudiana*. The study of polyploids may also have taxonomic significance, e.g. *Cochlearia*, *Symphytum* and *Mentha*. In some situations gene flow between polyploid levels takes place and may be in both directions. In other polyploid groups sexually sterile cytotypes may be maintained by vegetative reproduction, e.g. *Ranunculus ficaria* and *Holcus mollis* or by gametophytic apomixis. In some cases a polyploid and aneuploid series may be maintained in which all the components are fertile; this situation is discussed in relation to the genus *Erophila*. There has been considerable emphasis in the past in attempting to identify the genomic ancestors of allopolyploids; two British species, *Spartina anglica* and *Senecio cambrensis* are among the few examples of recent allopolyploids of known origin; in both cases formation has depended on the establishment of an introduced species as one parent.

The significance of B chromosomes is discussed; although a range of British species contain B chromosomes it has only been possible to investigate their adaptive significance in one case. Interchanges exist in British species either in the heterozygous form, e.g. in *Alopecurus* species, where they can be identified by their meiotic consequences, or in the homozygous form, e.g. *Epilobium* sect. *Epilobium* species, where they can be identified by meiotic analysis of hybrids.

The use of modern cytogenetic techniques including chromosome banding and *in-situ* hybridization and the investigation of characteristics such as infra-species variation in DNA amounts are discussed in relation to future cytogenetic investigations of the British flora.

Key words: British flora, cytogenetics, evolution, polyploidy.

INTRODUCTION

In Britain we have a long history of cytogenetic studies of the flora stretching back to the classical work of Blackburn & Heslop-Harrison (1921), who clearly showed, by their cytological investigations, that the *Rosa canina* group consists of unbalanced polyploids with a unique meiotic system, therefore explaining the reason for their complex pattern of variation. Only 3 years later Blackburn & Heslop-Harrison (1924) gave one of the first demonstrations of a natural polyploid series in their work on Salicaceae. From this period cytogenetic research on the British flora has both continued and expanded, particularly in the post-war period, this being not entirely unconnected with the stimulus given to the botanical research in Britain by the publication of the 'Flora of the British Isles' by Clapham, Tutin & Warburg in 1952. Among his wide botanical interests Roy Clapham was himself involved in cytogenetic research and published data on the cytology of *Juncus articulatus* and *J. acutiflorus* (with E. W. Timms) and worked, with B. L. Hancock, on *Galium palustre* which I will mention again later.

Today Britain has one of the best known floras from a cytological viewpoint.

Cranston (in Moore, 1978) has recently estimated that chromosome counts on British plants now exist for 42% of the flora, a figure only bettered by a few countries such as Iceland which has a 100% coverage of its more restricted flora. When, however, we come to look for more detailed information on the cytogenetics of the British flora the number of species covered adequately drops dramatically, but still comprises a substantial body of data. Another difficulty which has to be faced in the interpretation of the available data is that to understand our flora and its relationships it is often necessary to know the patterns of variation of species over their whole range. This is perhaps most clearly seen when considering the history and distribution of polyploids.

In this paper I shall try to show the value of cytogenetic variation in the interpretation of the British flora from a number of aspects. At the most obvious level cytogenetic studies, particularly chromosome counts and meiotic studies, have a significance and usefulness in enabling us to identify and distinguish taxa, which may then be considered separable at the specific, or for some at the infra-specific, level. Cytogenetic investigations may also lead us to understand how taxa have evolved or are still evolving and their relationships with one another. They may have a further significance in allowing us to trace the past immigration of taxa into this country and their subsequent history as part of the British flora. In order to consider some of the progress that has been made in these various areas it is most convenient to consider some examples in relation to the major cytogenetic processes.

POLYPLOIDY

Since the formation of polyploids is the most widespread cytogenetic process in flowering plants (Stebbins, 1971), it is not surprising that many cytogenetic studies on the British flora refer to polyploidy. They cover both a wide diversity of species and illuminate a range of aspects of their geographical distribution, their ecology and their evolutionary relationships.

Established polyploids and their phytogeographical significance

Polyploid series with two or more cytotypes in Britain. In a number of British genera and species, we have two or more closely related taxa at different ploidy levels, each with distinct ecology and distribution. One north temperate species to be studied earliest is *Empetrum nigrum*, in which Hagerup (1927) showed that Scandinavian plants are diploids ($2n = 26$) and dioecious, extending north to Iceland, while Greenland material is tetraploid ($2n = 52$) and hermaphrodite. In Britain the diploid *E. nigrum* ssp. *nigrum* is widespread in northern and western Britain, while the tetraploid *E. nigrum* ssp. *hermaphroditum* is restricted to the Lake District and Scottish Highlands where it replaces ssp. *nigrum* above c. 750 m (Bell & Tallis, 1973). Unfortunately, the physiological ecology of these two subspecies has not been investigated to elucidate the basis of this distribution pattern. A more subtle difference in distribution is shown by the diploid and tetraploid cytotypes of *Hippocrepis comosa* studied by Fearn (1972). Here the diploids ($2n = 14$) have a disjunct distribution being restricted to ungrazed and sparsely vegetated cliffs and rocky slopes on the older, harder limestones in the south and west of England and in Derbyshire, while the tetraploid cytotype ($2n = 28$) is common throughout the range of the species, being most common in grazed or formerly grazed chalk grassland. A number of the diploid's sites are noted for their rare plant assemblages and its distribution has strong similarities to that of *Helianthemum apenninum*, *H. canum* and *Aster linosyris*. Fearn (1972) has shown that the reproductive capacity of the diploids, in terms of inflorescence production, is much lower than that of tetraploids and also (Fearn, 1973) that prostrate plants, adapted to grazing pressure, have only evolved in the tetraploids. She has proposed that the explanation for their distribution patterns lies in their post-glacial history, suggesting that both cytotypes immigrated from the Continent, where they grow now, in the late glacial to early post-glacial periods. As forests became extensive they would have become restricted to scattered open sites, but subsequently as forest clearance took place, particularly on the southern chalklands, the tetraploid cytotype was able to colonize the grazed grasslands which developed, but the diploid cytotype, with a lower reproductive capacity and colonizing potential, remained restricted to ungrazed rock ledges. The situations in both *Empetrum* and *Hippocrepis* exemplify that identified by Ehrendorfer (1979) who has proposed that diploids are more common in stable habitats of permanent or climax communities, while their related polyploids are more common in disturbed or unsuccessful communities.

A more problematic situation exists in *Lamiastrum galeobdolon*, where Wegmüller (1971) discovered two localities in Lincolnshire for the diploid ssp. *galeobdolon* ($2n = 18$), which together with the widespread British tetraploid ssp. *montanum* ($2n = 36$) is present in both central and northern Europe. Subsequently only one further British site for ssp. *galeobdolon* has been found (Packham, 1983), in Kirkcudbrightshire, where there is doubt as to its native status. In such a situation it is very difficult to identify any differential ecological or historical factors within this generally distributed woodland species.

Unfortunately, there are still a number of species in this country containing polyploid complexes of which we know little more than of their existence. Regrettably two of these are groups studied by Roy Clapham, the *Galium palustre* group and the *Juncus articulatus* – *J. acutiflorus* group. Hancock (1942) and Clapham (1949) showed that *G. palustre* in this country contains diploid ($2n = 24$), tetraploid ($2n = 48$) and octoploid ($2n = 96$) cytotypes, which have both morphological and ecological differences; he found, however, that whereas the diploids and octoploids were clearly distinct morphologically, the tetraploid overlapped both in its characteristics. Subsequently Kliphuis (1974) has shown that the same three cytotypes exist in the Netherlands, while Teppner, Ehrendorfer & Puff (1976) have found that not only are these three cytotypes widespread in Europe, but that dodecaploids with $2n = 144$ also exist in Austria and Turkey. In contrast to Hancock (1942), Clapham (1949) and Kliphuis (1974) who favour treating the cytotypes as subspecies of *G. palustre*, Teppner *et al.* (1976) contend that although diploids and tetraploids are inseparable and are within *G. palustre*, the octoploids and dodecaploids are sufficiently distinct to form a separate species, *G. elongatum* Presl.

In *Juncus* (Timms & Clapham, 1940; Clapham, 1949) *J. articulatus* usually has $2n = 80$ chromosomes and *J. acutiflorus* has $2n = 40$ chromosomes; however, Timms & Clapham (1940) showed both that hybrids with $2n = 60$ exist and also plants with intermediate characteristics between these two species with $2n = 80$ chromosomes, so-called 'large 80' plants. Recently Zandee (1981) has made a large number of chromosome counts in the *J. articulatus* – *J. alpinus* group and has detected similar plants in samples from the Netherlands and Switzerland. Their intermediate morphology suggests that they may have a hybrid origin, possibly via unreduced gametes or they may be autotetraploids of *J. acutiflorus*. So far, however, no experimental work has been carried out.

Polyploid series with one cytotype in Britain. Often, of course, only one cytotype of a species is present in Britain, but cytotaxonomic studies over a wide area may provide valuable information on its history in Britain. In *Potentilla fruticosa* the British plants are tetraploid ($2n = 28$) and dioecious (ssp. *fruticosa*) as are those from Öland (Elkington, 1969). Using the character of dioecy, tetraploids have been identified in north Europe from the Baltic region and the Urals. In southern Europe, however, plants from the Pyrenees and Bulgaria are diploid ($2n = 14$) with hermaphrodite flowers [ssp. *floribunda* (Pursh) Elkington]. Material from the Maritime Alps and the Caucasus also has hermaphrodite flowers and is therefore from putative diploids. All north American material is also hermaphrodite and diploid. In Asia, clearly the centre of morphological variation, there are chromosome counts of $2n = 14$, 28 and 42. As far as the origin of the British populations is concerned, because the Baltic region was heavily glaciated in the last (Weichsel) glaciation, it seems likely that the tetraploids migrated along a northerly route from the east in the early post-glacial period extending to west Ireland. To the south, however, migration took place from the east of diploid stocks, either before or after the last glaciation. Clearly there is no direct phytogeographical connection, as has been suggested, between the Irish and south-west European populations.

Another example in which cytogenetic knowledge allows us to refine our ideas on plant migration is the study of Irish populations of *Sisyrinchium* (Ingram, 1967). These uniformly have chromosome counts of $2n = 88$ and are referable to *Sisyrinchium bermudiana*, which in eastern North America has an aneuploid series of chromosome numbers of $2n = 82$, 84, 88 and 90. Although not proven, it seems likely that *Sisyrinchium bermudiana* is native to Ireland. J. Heslop-Harrison (1953) has suggested that it was introduced by birds via Greenland as a 'half-way house'. Böcher (1966) has, however, discovered that Greenland populations of *Sisyrinchium* have $2n = 32$ and are clearly distinct from all north American species; he therefore described this taxon as a new species *Sisyrinchium groenlandicum* Böcher. Cytological evidence thus shows that Greenland is not involved in the establishment of *Sisyrinchium* in Ireland, although the origin of this and other species belonging to the 'American' element in the Irish flora remain problematical.

There are, of course, many other cases where we have cytological uniformity of a species in Britain, which forms part of a widely distributed polyploid complex of which we know little more than its general existence and distribution. In *Tripleurospermum inodorum* (Kay, 1969) the British populations form part of the western European distribution of a diploid cytotype ($2n = 18$), while the tetraploid cytotype ($2n = 36$) has a generally more eastern and continental distribution, except for one record from Oxfordshire, confined to the progeny of one individual, which Kay (l.c) suggests was introduced as a contaminant of crop-seed. In *Linum* (Ockenden, 1968, 1971), *Linum perenne* ssp. *anglicum*, to which all British populations belong, is tetraploid ($2n = 36$), while the majority of continental populations are diploid ($2n = 18$) except for the tetraploid ssp. *montanum* which is almost completely restricted to the Jura region. In *Myosotis alpestris* (Elkington, 1964), British populations are tetraploid ($2n = 48$), while a complex of diploid ($2n = 24$) and tetraploid forms are present in the Alps (Grau, 1964).

Polyploids and taxonomy

There is a number of genera in Britain where taxonomic problems have been clarified by cytogenetic study, which has also shed light on the evolution of their component species. This type of study is exemplified by the work on *Cochlearia* carried out by Gill (1965, 1971a, b, 1973, 1976) and Gill, McAllister & Fearn (1978). Chromosome numbers in *Cochlearia* fall into two series with basic numbers of $x = 6$ and 7 (see Fig. 1). Inland British populations have $2n = 12$, 24 or 26, except for one doubtful count of $2n = 14$. Although diploid plants with $2n = 12$ and tetraploids with $2n = 24$ may be extremely difficult to separate morphologically, owing to considerable phenotypic plasticity, it is clear that tetraploids are identifiable with the common coastal species *C. officinalis*, while the diploids may be referred to *C. pyrenaica*. This latter species is almost exclusively confined to heavy metal contaminated sites in north-west Europe. Plants with $2n = 26$ are distinctive in cytology, morphology and distribution, are confined to high altitude sites in Scotland and are identifiable as *C. micacea* which is, perhaps, a British endemic. The old count of $2n = 14$ is of interest because, if confirmed, it would form the southern limit of a cytotype known otherwise from the arctic and subarctic area south to Iceland (Gill, 1971a). It has been identified by Gairdner (1939) as *C. scotica*. Gill (1971a, 1973) has proposed that plants with $2n = 14$ evolved from a $2n = 12$ ancestor and are primary tetrasomics, while *C. officinalis* is an autotetraploid of the diploid *C. alpina*, and *C. micacea* ($2n = 26$) is a primary tetrasomic of a $2n = 24$ ancestor.

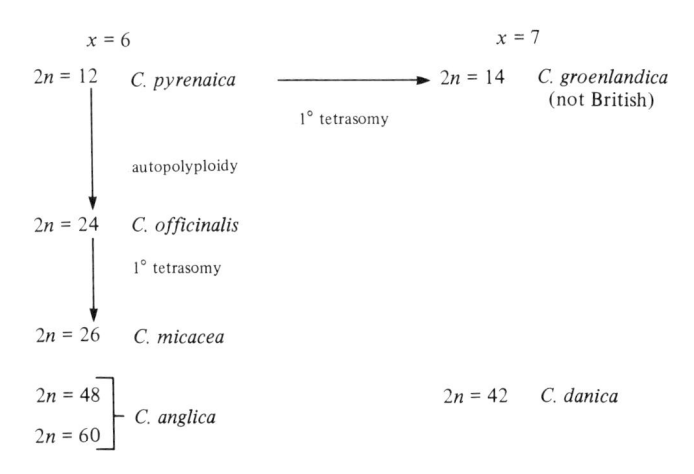

```
        x = 6                                              x = 7

2n = 12    C. pyrenaica   ──────────────►   2n = 14    C. groenlandica
   │                                                    (not British)
   │                          1° tetrasomy
   │
   │       autopolyploidy
   ▼
2n = 24    C. officinalis
   │
   │       1° tetrasomy
   ▼
2n = 26    C. micacea

2n = 48 ⎤
        ⎬  C. anglica                       2n = 42    C. danica
2n = 60 ⎦
```

Fig. 1. Chromosome numbers and relationships in British *Cochlearia* species. (Data from Gill, 1965, 1971a, 1973, 1976.)

In a number of cases although we have well established species in a polyploid series hybridization has resulted in a confused taxonomic situation only clarified by careful cytogenetic studies. In *Symphytum*, *Symphytum × uplandicum* Nyman was originally thought to consist of F_1 and backcross hybrids derived from crosses between *Symphytum asperum* and *Symphytum officinale* for which chromosome numbers of $2n = 36$ and $2n = 42$, respectively, were available (Perring, 1969). Cytological examination of British plants (Gadella, Kliphuis & Perring, 1974) shows a different and more complex situation. *Symphytum officinale* has both diploid ($2n = 24$) and tetraploid cytotypes ($2n = 48$) and *Symphytum asperum*, a rare relic of cultivation in Britain has $2n = 32$. *Symphytum × uplandicum*, however, exists as two cytotypes with $2n = 36$ and 40. The $2n = 36$ forms are thought to originate from hybridization of *Symphytum asperum* with a $2n = 40$ cytotype of *Symphytum officinale*, unknown in Britain, but common in fens in the Netherlands (Gadella & Kliphuis, 1971). The $2n = 40$ hybrids are regarded as being derived from a cross of *Symphytum asperum* with the $2n = 48$ cytotype of *Symphytum officinale*. Both *Symphytum × uplandicum* types were almost certainly originally introduced.

One of the most complex situations with regard to hybrids exists in *Mentha* sect. *Mentha*, most recently studied by Harley (1975) and Harley & Brighton (1977). Here hybrids are readily formed between species at various polyploid levels and may exist in the absence of their parents owing both to their continuing spread by vegetative reproduction, through their stoloniferous habit, and because of their widespread cultivation. The species in *Mentha* sect. *Mentha* form a polyploid series from diploid to octoploid levels with a basic number of $x = 12$; hybrids may be formed between species at the same and different ploidy levels. Cytogenetic studies have been particularly useful in elucidating the ancestry of some of these hybrids. In some cases the results confirm morphological interpretations of hybrids as being of straightforward F_1 origin; this is so for *M. × gentilis* which has $2n = 60$ chromosomes (except for one aneuploid count of $2n = 61$), and is thus consistent with its assumed origin as a hybrid of *M. arvensis* ($2n = 72$) and *M. spicata* ($2n = 48$). Similarly *M. × piperita* clones, some of which are used as a source of peppermint oil, nearly all have $2n = 72$ which is consistent with their being F_1 hybrids between *M. aquatica* ($2n = 96$) and *M. spicata* ($2n = 48$). In *M. × villosa* all counts are of $2n = 36$ chromosomes; that *M. suaveolens* is clearly one parent is indicated from the morphology of the hybrid and pubescent forms were originally thought to involve *M. longifolia* (L.) Huds. as the other parent. This is unlikely, however, because it has $2n = 24$ chromosomes and the alternative of *M. spicata* with $2n = 48$ is more plausible. In another case Harley & Brighton (1977) found a putative hybrid of *M. aquatica* ($2n = 96$) and *M. suaveolens* ($2n = 24$) growing with its assumed parents, but whereas a hybrid of this parentage would be expected to have $2n = 60$ chromosomes, the plants had $2n = 72$. However, a form of *M. spicata* ($2n = 48$), growing in the same area and having a similar appearance to *M. suaveolens*, could be one of the parents since an F_1 hybrid of this and *M. aquatica* would be expected to have $2n = 72$. Clearly here cytogenetic investigations complement careful morphological comparisons in establishing the origins of hybrids in this complex genus.

Gene flow through polyploids

Although the formation of polyploids is conventionally thought of as resulting in a sterility barrier between diploids and polyploids it has become clear from recent investigations that gene flow may take place between ploidy levels. This may have considerable significance for the further evolution of these groups and for their taxonomic interpretation. In the cases investigated, with the possible exception of *Betula*, mentioned below, gene flow is dependent on the presence of partially fertile and cytologically intermediate F_1 plants. The range of possibilities is apparent from Lord & Richards' (1977) study of a mixed population of *Dactylorhiza fuchsii* ($2x, 2n = 40$) and *D. purpurella* ($4x, 2n = 80$) which had formed a hybrid swarm. Meiotic examination of individuals showed that as well as euploid putative hybrids with $2n = 60$ and variable morphology, aneuploids with $2n = 44$, 48, 52 and 72 were also present. The presence of aneuploid plants with chromosome numbers both above and below triploid $2n = 60$ counts suggests that backcrossing to both parents is taking place and that the triploids must therefore have partial fertility. The considerable variation among triploids suggests that they are not all of F_1 origin. Observations of their meiosis showed that in some cells 20 bivalents and 20 univalents were formed; if bivalents segregated regularly, together with movement of all univalents to one or the other pole, gametes with $n = 20$ or 40 could be produced. Fertilization of such gametes from the parental species could therefore give both new triploids and hybrid derivatives with $2n = 40$ and 80.

In *Cochlearia*, Fearn (1977) has studied an introgressed population of *C. officinalis* ($4x$, $2n = 24$) and *C. danica* ($6x$, $2n = 42$). Here, intermediates had haploid chromosome numbers of $n = 12-17$, indicating that backcrossing is taking place between pentaploid F_1 plants and *C. officinalis*, with progressive loss of unpaired chromosomes at meiosis, probably originating from *C. danica*, so that the net result is the introgression of *C. danica* genes into the tetraploid *C. officinalis*.

Another mechanism has been shown to exist in *Senecio* by Crisp & Jones (1978) who studied *Senecio squalidus* ($2x$, $2n = 20$), *Senecio viscosus* ($4x$, $2n = 40$) and their F_1 hybrid, *Senecio × londinensis* Lousley ($3x$, $2n = 30$). Here, examination of F_2 and further generations of progeny from *Senecio × londinensis* showed that production of pentaploids with $2n = 50$ and aneuploids with $2n = 47$, 48 and 49 took place, to be followed in later generations by loss of chromosomes and return to a near tetraploid state. Thus gene flow appears to be from diploid to tetraploid levels with both triploid and pentaploid intermediates being involved.

Gene flow in the opposite direction has been proposed by Yeo (1956) to account for variable populations of the diploid *Euphrasia anglica* Pugsl. ($2n = 22$); here gene flow is proposed via triploid hybrids with *E. micrantha* Rehb. ($4x$, $2n = 44$) which produce viable gametes with $n = 11$ which are then fertilized by *E. anglica* gametes so that gene flow is from tetraploid to diploid levels.

A much more intractable situation exists in *Betula* where *B. pendula* ($2x$, $2n = 28$) and *B. pubescens* ($4x$, $2n = 56$) often co-exist in woodlands which then contain trees of varying morphology, perplexing to interpret. Cytogenetic investigations have given varying results. In Scotland, Brown & Al-Dawoody (1977) have studied a population of trees with modal chromosome numbers of $2n = 28$, 42 and 56 plus several aneuploids, chromosome counts of which range in total from $2n = 37$ to 56. The $2n = 42$ individuals are indistinguishable morphologically from those with $2n = 56$ and they form a variable group, generally identified as belonging to *B. pubescens*. In contrast, in East Anglia, (Gill & Davy, 1983) a similar mixed and variable population had only individuals with $2n = 28$ or 56 chromosomes. Studies by Brown & Al-Dawoody (1979) on meiosis of Scottish trees show relative regularity in division within the individuals with $2n = 28$ and 56, whereas trees with $2n = 42$ showed many irregularities at division, particularly chromosome clumping; nevertheless approx. 40 to 50% of observed divisions lead to gametes with 21 chromosomes. They suggest that this behaviour is not consistent with the hypothesis that trees with $2n = 42$ are triploid, but are more likely to be hexaploid, with individuals having $2n = 28$ and 56 being tetraploid and octoploid, respectively. Additionally, phenetic data do not suggest a hybrid origin for the $2n = 42$ individuals which may represent aneuploid *B. pubescens* types. In East Anglia Gill & Davy (1983) proposed that some of the $2n = 56$ trees may be the result of hybridization with *B. pendula* via unreduced gametes. This would then be a mechanism for gene flow between the two chromosome levels. Here more cytogenetic investigations are needed on what is, unfortunately, very difficult cytological material.

Established polyploid complexes

There are a number of polyploid groups, often regarded as individual species, in which several cytotypes may be permanent components of the British flora. In some cases one or more cytotypes may be quite sterile, but are maintained through some form of apomixis.

One mechanism is through some form of vegetative reproduction. In *Ranunculus ficaria* widespread sampling (Gill et al., 1972) has shown that British populations include diploid ($2n = 16$), triploid ($2n = 24$) and tetraploid ($2n = 32$) plants and either comprise a single cytotype or are mixed. Nine of the populations, 4·1%, contained triploids which do not produce seed; although some aerial bulbils are produced the main reproductive mechanism of these individuals is through the numerous and easily detached tubers produced by all *R. ficaria* plants. Marchant & Brighton (1974) have proposed that if diploids, which require cross pollination, have a scattered distribution among an excess of tetraploids, then new triploids may be produced with considerable frequency. As far as the distribution of the predominant diploid and tetraploid races is concerned Nicholson (1983) has suggested, from studies in S.E. Yorkshire, that they are ecologically distinct, the diploid being more light-tolerant than the tetraploid, the present mixed distribution in many parts of Britain being due to long term habitat disturbance.

In *Holcus mollis* (Jones, 1958) the most common chromosome number in British plants is $2n = 35$, a pentaploid level, and the same situation has been shown to exist in the Netherlands (Zandee & Scheenen, 1981). Although this cytotype is capable of producing some seed, most of these progeny are aneuploid and ill-adapted to survive under natural conditions. Effective reproduction of the pentaploid is by rhizomes and its success depends both on its high vegetative vigour and its wide ecological adaptability in comparison with the other three British cytotypes ($2n = 28$, 42 and 49). The second most common race is the tetraploid ($2n = 28$) which, although fertile, seems to colonize mainly by rhizomes and is thus in direct competition, in this respect, with the pentaploid. The vigour of the pentaploid is correlated with its hybrid nature, Jones (1958), Jones & Carroll (1962) and Carroll & Jones (1962) having shown that it is derived from a back-cross between tetraploid *H. mollis* and an unreduced gamete from the triploid hybrid of the diploid *H. lanatus* ($2n = 14$) and tetraploid *H. mollis* ($2n = 28$).

Polyploid complexes are also common among groups reproducing by gametophytic apomixis. Examples include *Pontentilla tabernaemontani* ($6x$, $2n = 42$; $7x$, $2n = 49$; $8x$, $2n = 56$; $9x$, $2n = 63$; $10x$, $2n = 70$) (Smith, 1963a; Smith, Bozman & Walters, 1971), *Potentilla crantzii* ($6x$, $2n = 42$; $7x$, $2n = 49$; $9x$, $2n = 63$) (Smith, 1963b; Smith, Bozman & Walters, 1971), the *Rubus fruticosus* aggregate ($3x$, $2n = 21$ to $7x$, $2n = 49$) (Y. Heslop-Harrison, 1953), the genus *Sorbus* ($3x$, $2n = 51$; $4x$, $2n = 68$) (Warburg, 1962) and many microspecies in *Taraxacum* ($3x$, $2n = 24$) (Richards, 1973).

Some polyploid complexes contain a range of euploid and aneuploid cytotypes which, however, are fertile and sexually reproducing. One excellent example is the genus *Erophila* now being studied by S. A. Filfilan at Sheffield. The taxonomy of this inbreeding and extremely plastic group was first treated in detail by Jordan (1852, 1864) who described a large number of new species, the so-called 'Jordanons' of later taxonomists (see Stace, 1980); these were later disregarded and the genus has received a different treatment in almost every flora. A major landmark was provided by Winge (1940) who carried out a detailed survey of north European populations, showing the existence of a polyploid and aneuploid complex with chromosome numbers of $2n = 14-64$. Winge also gave a taxonomic treatment which, unfortunately, took no account of previous taxonomy and has thus been largely ignored. Filfilan's studies have shown that in Britain the cytology parallels that of Continental populations. Diploid populations with $2n = 14$ chromosomes are distinctive in morphology and can be referred to the Jordan species *E. majuscula*, corresponding to *E. simplex* Winge. Plants with $2n = 30, 32, 34, 36, 40, 42$ and 44, of which $2n = 36$ is the most common British cytotype, have a number of morphological similarities and are referable to *E. verna sensu stricto*, corresponding to *E. duplex* Winge. Plants with $2n = 48, 52, 54$ and 56 (and $2n = 58$ and 64 on the Continent) can be distinguished as the Jordan species *E. glabrescens*, corresponding to *E. quadriplex* Winge. This cytotaxonomic treatment has shown both the utility of using previously disregarded morphological characters for specific delimitation and that the previous emphasis on using capsule characters was misplaced. Although we now have a sound basis for a taxonomic treatment, the problems of the relationships and evolution of these cytotypes remain to be answered. Studies by Filfilan of the karyotypes show the diploids to be uniform. Expected tetraploids with $2n = 28$ have never been reported and cytotypes with $2n = 32-44$ chromosomes have a distinctive basic karyotype pattern not easily derived from the diploid karyotype; similarly the karyotypes of the $2n = 48-56$ cytotypes are similar to one another, but not to the other two karyotype patterns. The probable origin of some cytotypes is suggested by crosses made between cytotypes; thus a cross between plants with $2n = 48$ and 32 gave F_1 plants with $2n = 40$ showing irregularities at meiosis. An F_2 generation produced by selfing gave plants with chromosome numbers of $2n = 44, 46, 48, 52$ and 56. Stabilization of chromosome number and meiotic behaviour in later generations would therefore provide a mechanism for the production of some cytotypes. In contrast a cross between plants with $2n = 48$ and 52, having similar karyotypes, gave an F_1 generation with $2n = 50$, having a regular meiosis with the formation of 25 bivalents at first metaphase, and an F_2 generation of plants also uniformly having $2n = 50$.

Evolution of polyploids
    A considerable number of British genera comprise a polyploid series with one or more species at each ploidy level. Cytogenetic investigations in these groups has often focused on elucidating the evolutionary relationships between species, particularly with respect to their genomic composition. In some cases this approach has been relatively successful. For example in *Viola* Moore & Harvey (1961) proposed that the diploid *V. persicifolia* (*V. stagnina*) ($2n = 20$) is one of the putative parents of the allotetraploid *V. canina* ($2n = 40$), while *V. reichenbachiana* is one parent of the allotetraploid *V. riviniana* ($2n = 58$) which also includes one unknown genome (B) in common with *V. canina* (see Fig. 2). They also showed that the allohexaploid *V. lactea* ($2n = 58$) also includes the two genomes (B, C) present in *V. canina*, plus D, the ancestor of which has also not been identified. This relatively clear cut result is unfortunately not common in such studies. A more common situation is that described by Humphreys (1975) in *Papaver*. Here, although hybrids were successfully produced from crosses involving *P. rhoeas* (2x,

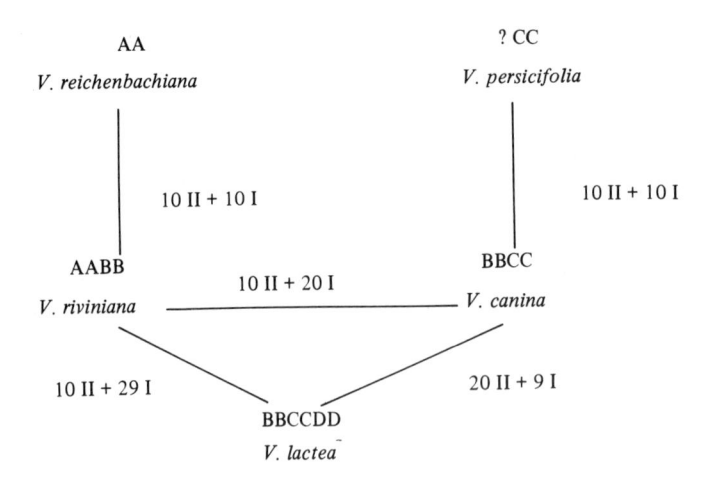

Fig. 2. Possible genome relationships and meiotic pairing between British species of *Viola* sect. Rostratae (from Moore & Harvey, 1961).

$2n = 14$), *P. lecoqii* ($4x$, $2n = 28$) and *P. dubium* ($6x$, $2n = 42$) interpretation of possible genomic relationships through meiotic analyses were obscured by genetic control of homologous pairing and by homoeologous pairing. As Stace (1975) has pointed out, genomic analysis has only provided a satisfactory ancestry in a minority of groups owing to homoeologous and non-homologous pairing, genetic control of pairing and the further evolution of the original ancestral species. Even when there is good evidence for the identification of diploid ancestors, as in the presumed allotetraploid *Poa annua* ($2n = 28$) (Tutin, 1957), further research (Ellis, Calder & Lee, 1970), in this case of the discovery of diploid populations ($2n = 14$) from Australia with regular bivalent formation, may indicate that the situation is more complex than previously recognized.

Two British polyploids *Spartina anglica* and *Senecio cambrensis* have a special place in cytogenetic studies since they are among the very few examples of new allopolyploids which have evolved in the last hundred years in known circumstances. In *Spartina*, Marchant (1967, 1968) has shown that the sterile F_1 hybrid between the native *Spartina maritima* ($2n = 60$) and the north American *Spartina alterniflora* ($2n = 62$), only known to have been introduced into Southampton Water in Britain, was first collected from there in 1870, approximately 50 years after the first introduction of *Spartina alterniflora*. He believes that the amphidiploid and fertile *Spartina anglica* arose directly there from the F_1 hybrid in approx. 1890. The parental species are both hexaploid so that the amphidiploid is approximately $12x$ with chromosome numbers of $2n = 120$, 122 and 124. Even though formation of *Spartina anglica* took place relatively recently it is still uncertain how it took place, although Harlan & De Wet (1975) have pointed out that since amphidiploids with several chromosome numbers exist, an origin through meiotic processes seems most likely, rather than by somatic doubling. Unfortunately Marchant (1968) was unable to produce an F_1 hybrid experimentally or to produce amphidiploids by colchicine treatment of natural hybrids.

Senecio cambrensis ($2n = 60$) was described by Rosser (1955), from plants collected in N. Wales; she also recorded that Harland and Jackson had produced similar plants with $2n = 60$ chromosomes from colchicine treatment of a synthetic triploid hybrid of *Senecio squalidus* ($2x$, $2n = 20$) and *Senecio vulgaris* ($4x$, $2n = 40$).

Synthetic allohexaploids have also been produced by Weir & Ingram (1980) using the same technique, and they also identified an allohexaploid produced spontaneously by an F_1 hybrid, although it is not known if this resulted from fusion of unreduced gametes or whether chromosome doubling had occurred during early development. The synthetic allohexaploids are of interest in having high fertility associated with regular meiosis; despite this *Senecio cambrensis* has, apparently, been unable to spread more than 11 km from its assumed site of origin in Wales except for one site 40 km away, at Colwyn Bay, Clwyd (Stace, 1977). Recently Abbott, Ingram & Noltie (1983) have identified *Senecio cambrensis* in the Edinburgh area; it is not clear if this is the result of long distance dispersal from Wales or an independent origin. It may be relevant, as Abbott *et al.* (1983) point out, that at the time that *Senecio cambrensis* originated in Wales *Senecio squalidus* was colonizing Wales, while its colonizing front has now reached the central industrial belt of Scotland and indeed F_1 hybrids of *Senecio squalidus* × *Senecio vulgaris* are known from Edinburgh.

Spartina anglica and *Senecio cambrensis* provide an interesting comparison in that their production has depended on the presence of an introduced species; in both cases evolution has taken place within a relatively short time of that introduction and has almost certainly involved an intermediate and sterile F_1 hybrid. Both have been dependent for their continued existence and colonization on the existence of open habitats; bare mud flats in the case of *Spartina anglica* and roadsides and waste ground for *Senecio cambrensis*. However, whereas *Spartina anglica* spread by rapid natural colonization from its original locality (Marchant, 1967), probably mainly by vegetative spread of rhizomes, *Senecio cambrensis*, even though highly fertile, has colonized relatively slowly from its original site. These examples may suggest that under natural or semi-natural conditions, where relatively large numbers of individuals may be involved, potential allopolyploids will evolve relatively quickly after their parental species come into contact but established stable floras are unlikely to generate new allopolyploids.

These examples may be contrasted with the formation of new autopolyploids which may continue to be produced sporadically by established species. This process has recently been studied in *Alopecurus bulbosus* (Sieber & Murray, 1980) a rare British species at the northern edge of its distribution. Two of three populations studied contained triploids ($2n = 21$) and tetraploids ($2n = 28$) as well as the widely distributed diploids ($2n = 14$). Observations showed that the probable origin of the tetraploids was by somatic doubling, with triploids being formed by subsequent hybridization between diploids and tetraploids; both triploids and tetraploids are thus autopolyploids. This type of phenomenon may not be uncommon, but sampling of populations is often inadequate and it remains to be seen if such autopolyploids can successfully establish themselves in the long term.

B CHROMOSOMES

The majority of references to the presence of B chromosomes represent observations on individuals or small numbers of plants (Jones, 1975). Even when extensive studies have been carried out the frequency of B chromosomes has often been so low that little more than their frequency in individuals and populations has been established. Bosemark (1956) found in *Festuca pratensis* that only 4·4% of a total of 1654 plants sampled from populations across southern England contained B chromosomes, in contrast to a frequency of approximately 20% in samples from

Swedish populations. Similarly Sieber & Murray (1981) found 1 or 2 B chromosomes in a minority of *Alopecurus myosuroides* plants in two out of eight British populations which were sampled. In other species, e.g. *Alopecurus pratensis* (Sieber & Murray, 1981) and *Briza media* (Murray, 1976a) B chromosomes are apparently absent from Britain, although present in continental populations. The significance of these differences is not clear, although Bosemark (1956) states that British plants of *Festuca pratensis* with B chromosomes come from the most recently established populations so that they may be introductions.

Even when relatively high proportions of plants contain B chromosomes it is often extremely difficult to find any significance or pattern in their distribution. In *Ranunculus ficaria* Gill et al. (1972) and Marchant & Brighton (1974) have shown that individuals possessing B chromosomes are widespread and may be common in populations, but are confined to diploid plants. In one mixed population of diploids, triploids and tetraploids, Marchant & Brighton (1974) found that 57% of the diploids contained one to eight B chromosomes, but although they were able to show a preferential distribution pattern, with B chromosome plants being most common in a copse area, they were unable to find any explanation for it. In *Allium schoenoprasum* (Bougourd & Parker, 1975), the number of B chromosomes per cell varies geographically in populations along the River Wye. The upper populations have no B chromosomes, but those downstream contain plants with B chromosomes, rising to a maximum of 65% of such plants in a population. Intensive sampling (Bougourd & Parker, 1979a) has shown that there is an abrupt change in frequency from zero to 54% of plants containing B chromosomes in a distance of 550 m. There are nine morphologically distinct types of B chromosome, all euchromatic, which can be derived by processes of centric misdivision, deletion and centric shift from a basic telocentric type which is by far the most common category. Individual plants may contain up to 18 B chromosomes, although plants with more than six B chromosomes are rare. Despite the high frequencies of B chromosome plants at some sites Bougourd & Parker (1979b) have shown that the B chromosomes have deleterious effects on vigour and fertility, although there is little difference between plants with zero, one, two and three B chromosomes. Holmes & Bougourd (1983) have attempted experimentally to investigate the effects of stress in terms of sowing density and water regime on seedlings containing a range of B chromosomes, but no significant differences were detected. Along the River Wye, however, there is evidence that seedlings with B chromosomes are at a selective advantage in terms of survival, although the nature of this advantage is not clear (Holmes & Bougourd, pers. comm.). There is therefore a possible adaptive explanation for the survival of plants with B chromosomes. There is no evidence for inherent mechanisms for the accumulation of B chromosomes in *A. schoenoprasum* and often slight loss occurs (Bougourd & Parker, 1979b) which also suggests that their common presence in these populations is related to some type of selective advantage.

An interesting situation, which indicates one type of origin of B chromosomes, is provided by the study, already mentioned, of an introgressed *Cochlearia* population by Fearn (1977). Here, the putative hybrids between *C. officinalis* ($4x$, $2n = 24$) and *C. danica* ($6x$, $2n = 42$) have haploid chromosome numbers of $n = 12-17$. Meiotic studies of the aneuploids showed variable, but low numbers of univalents and high proportions of multivalents, although these irregularities do not, apparently, greatly affect the viability of the pollen, as judged by its stainability. Backcrossing seems to be mainly to *C. officinalis*, unpaired chromosomes, probably belonging to *C. danica*, being gradually eliminated at meiosis. Residual unpaired chromosomes, however, may form the B chromosomes previously recorded in *C. officinalis* and found here in one plant of this species.

A measure of the problems to be faced in studying B chromosomes is evidenced by the situation in *Silene maritima*. Cobon & Murray (1983) have found that plants with B chromosomes are common in Norfolk populations, but are found nowhere else in Britain; the number of B chromosomes not only varies widely between individuals, but cells of the same plant may vary to the extent of having between one and 14 B chromosomes in root cells and one to five B chromosomes in pollen mother cells. Another problem, illustrated by *Caltha palustris* (Kootin-Sanwu & Woodell, 1969), is that progeny of an individual may vary considerably from the parental constitution; thus progeny of a plant with $2n = 56$ and lacking B chromosomes, had $2n = 55$, 56 and 57 A chromosomes and up to three B chromosomes.

INTERCHANGES

The most common and best studied structural alterations to chromosomes in plants of the British flora are interchanges (reciprocal translocations) which have been shown to exist in both heterozygous and homozygous forms in several species. In *Alopecurus*, Sieber & Murray (1981) have shown that the perennial species *A. bulbosus* and *A. myosuroides* and the annual *A. aequalis*, all with $2n = 14$, have populations which contain a proportion of interchange heterozygotes which are identifiable by the presence of one quadrivalent and five bivalents at first meiotic metaphase. The interchanges result in a variable reduction in fertility and, although this may not be so impotant in the two perennial species, its continued existence in the annual *A. aequalis* may be due to a variety of factors including population size and heterozygous advantage. A similar situation has been found in *Briza media* (Murray, 1976b) where some populations contain individuals with heterozygous interchanges.

The formation of interchanges may be an integral part of the evolutionary history of species groups and this has been well documented by Seavey & Raven (1977a, b) in their studies of *Epilobium* sect. *Epilobium* to which the British species belong. Here a wide crossing programme and meiotic analysis of the F_1 hybrids have shown that many of the species, which are all diploid with $2n = 18$, differ from one another by homozygous interchanges; species may then be grouped by their chromosomal arrangements as shown here for British species (Fig. 3). The arrangement BB, common in Eurasia, Africa and also present in the Americas appears to be a basic one, AA, CC and DD each differing from it by a single

AA	*E. alsinefolium*
BB	*E. hirsutum*
	E. lanceolatum
	E. montanum
	E. obscurum
	E. palustre
	E. parviflorum
	E. roseum
	E. tetragonom
CC	*E. anagallidifolium*

Fig. 3. British species of *Epilobium* sect. *Epilobium* with chromosome arrangements. (Data from Seavey & Raven, 1977a, b.)

interchange (Seavey & Raven, 1977b). The AA arrangement occurs in North and South America and at least three European species, including *E. anagallidifolium*; the CC arrangement, which differs from AA by two interchanges, characterizes a circumboreal group of species including *E. alsinefolium*. The results suggest that these changes are long-established, considerable species differentiation having taken place in several of these groups subsequent to interchange formation.

In other genera interchanges may form part of the genetic changes which have taken place in the evolution of individual species. Thus in *Hypochaeris* (Parker, 1975), *H. glabra* ($2n = 10$) and *H. radicata* ($2n = 8$) differ from one another by at least three interchanges. In this species pair, the differences between them in aneuploid chromosomes are not closely related to the formation of unequal interchanges in *H. glabra*, followed by chromosomal loss as has, for example, been implicated in the evolution of species in other genera, such as *Crepis* (Sherman, 1945) and *Haplopappus* (Jackson, 1962). Both species may, however, have been derived by this process from a $2n = 10$ ancestor (Parker, 1975).

Conclusions

I have examined only a selection of examples here, but they are sufficient to show that cytogenetic studies have a clear significance in the study of the British flora. There is, however, much to be done in continuing this work. Less than 50% of our flora has a chromosome count of even one individual from a native locality and, although not all species will be equally illuminated by such studies, in many genera problems of their taxonomy, history and relationships will be resolved only be detailed cytogenetic work.

I should make it clear, however, that not only do we need to go further down the same path as before, but there are a range of cytogenetic phenomena hardly studied so far in the British flora. For example we known almost nothing about cytogenetic variation at the population level. The older studies of heterochromatin distribution, as revealed by cold treatment, in *Trillium* populations (cf. Fukuda & Channell, 1975) show the wealth of information which may be gained. We now have newer techniques of much wider application utilizing Giemsa stain (Vosa, 1975) and fluorescent dyes (Schweizer, 1979) which allow resolution of considerable variation along individual chromosomes to give characteristic banding patterns. Many of the studies on plants using these techniques are concerned with investigations of species relationships e.g. in *Triticum* (Gill & Kimber, 1974), *Allium* (Vosa, 1976a, b; Badr & Elkington, 1977), *Scilla* (Greilhuber, Deumling & Speta, 1981) and there have been only relatively few attempts to use them for population analysis, e.g. *Scilla sibirica* Haw. (Vosa, 1973a) and *Tulbaghia leucantha* (Vosa, 1973b). Investigations using these techniques on suitable species may reveal unexpected population differentiation. Even more recently investigations have commenced on species relationships by *in-situ* hybridization, both of satellite DNAs, as in the *Scilla sibirica* group investigated by Deumling & Greilhuber (1982), and of cloned sequence families, e.g. in *Secale cereale* (Appels et al., 1981, Jones & Flavell, 1982).

It has also recently become apparent that variation in the amount of DNA may exist within species, quite independent of any variation in chromosome number. In *Scilla autumnalis* (Ainsworth, Parker & Horton, 1983) two diploid races ($2n = 14$) differ by 70% in their DNA content, although having essentially similar karyotypes in terms of relative size, and centromere and nucleolar positions; one (AA), with a 4C DNA content of 30·6 pg, is known only from W. Portugal, while the other (B^7B^7), with a 4C DNA value of 17·9 pg, grows in the northern Mediterranean region. In the British Isles, populations are either of the autotetraploid $B^7B^7B^7B^7$, present in southern England, or of the autoallohexaploid $AAB^7B^7B^7B^7$, found in W. Cornwall and the Channel Islands. Mowforth & Grime (in Grime, 1983) have shown that seedling populations of *Poa annua* grown from seed collected from individuals in a single pasture population differ in mean DNA content between 2·7 and 4·9 pg per 2C nucleus, although it is not clear if similar variation exists in natural populations, which have been subject to selection at the seedling stage. In this case DNA content is correlated with rates of dry matter production so that selection of particular DNA genotypes is likely. Climatic and ecological selection of populations in relation to DNA content has been shown in *Microseris douglasii* (DC.) Sch.-Bip. (Compositae) in California by Price, Chambers & Bachmann (1981), who found a DNA variation of more than 20%, with higher DNA populations being restricted to more mesic sites. Clearly therefore we are on the threshold of a quite new era of cytogenetic investigations which will undoubtedly reveal as much about the evolution and history of the British flora as past studies have done.

REFERENCES

ABBOTT, R. J., INGRAM, R., & NOLTIE, H. J. (1983). Discovery of *Senecio cambrensis* Rosser in Edinburgh. *Watsonia*, **14**, 407–408.

AINSWORTH, C. C., PARKER, J. S. & HORTON, D. M. (1983). Chromosome variation and evolution in *Scilla autumnalis*. In: *Kew Chromosome Conference*, vol. II (Ed. by P. E. Brandham & M. D. Bennett), pp. 261–268. G. Allen & Unwin, London.

APPELS, R., DENNIS, E. S., SMYTH, D. R. & PEACOCK, W. J. (1981). Two repeated DNA sequences from the heterochromatic regions of Rye (*Secale cereale*) chromosomes. *Chromosoma*, **84**, 265–277.

BADR, A. & ELKINGTON, T. T. (1977). Variation of giemsa C-band and fluorochrome banded karyotypes and relationships in *Allium* subgen. *Molium*. *Plant Systematics and Evolution*, **128**, 23–35.

BELL, J. N. B. & TALLIS, J. H. (1973). *Empetrum nigrum* L. Biological Flora of the British Isles. *Journal of Ecology*, **61**, 289–305.

BLACKBURN, K. B. & HESLOP-HARRISON, J. W. (1921). The status of the British rose forms as determined by their cytological behaviour. *Annals of Botany*, **35**, 159–188.

BLACKBURN, K. B. & HESLOP-HARRISON, J. W. (1924). A preliminary account of the chromosomes and chromosome behaviour in the Salicaceae. *Annals of Botany*, **38**, 361–378.

BÖCHER, T. W. (1966). Experimental and cytological studies on plant species X *Sisyrinchium* with special reference to the Greenland representative. *Botanisk Tidsskrift*, **61**, 273–290.

BOSEMARK, N. O. (1956). On accessory chromosomes in *Festuca pratensis*. III. Frequency and geographical distribution of plants with accessory chromosomes. *Hereditas*, **42**, 189–210.

BOUGOURD, S. & PARKER, J. S. (1975). The B-chromosome system of *Allium schoenoprasum* I. B-distribution. *Chromosoma*, **53**, 273–282.

BOUGOURD, S. M. & PARKER, J. S. (1979a). The B-chromosome system of *Allium schoenoprasum* III. An abrupt change in B-frequency. *Chromosoma*, **75**, 385–392.

BOUGOURD, S. M. & PARKER, J. S. (1979b). The B-chromosome system of *Allium schoenoprasum* II. Stability, inheritance and phenotypic effects. *Chromosoma*, **75**, 369–383.

BROWN, I. R. & AL-DAWOODY, D. M. (1977). Cytotype diversity in a population of *Betula alba* L. *The New Phytologist*, **79**, 441–453.

BROWN, I. R. & AL-DAWOODY, D. (1979). Observations on meiosis in three cytotypes of *Betula alba* L. *The New Phytologist*, **83**, 801–811.

BROWN, I. R., KENNEDY, D. & WILLIAMS, D. A. (1982). The occurrence of natural hybrids between *Betula pendula* Roth. and *B. pubescens* Ehrh. *Watsonia*, **14**, 133–145.

CARROLL, C. P. & JONES, K. (1962). Cytotaxonomic studies in *Holcus*. III. A morphological study of the triploid F_1 hybrid between *Holcus lanatus* L. and *H. mollis* L. *The New Phytologist*, **61**, 72–84.

CLAPHAM, A. R. (1949). Taxonomic problems in *Galium* and *Juncus*. In: *British Flowering Plants and Modern Systematic Methods* (Ed. by A. J. Wilmott), pp. 72–74. Botanical Society of the British Isles.

CLAPHAM, R. R., TUTIN, T. G. & WARBURG, E. F. (1952). *Flora of the British Isles*. Cambridge University Press, Cambridge.

COBON, A. M. & MURRAY, B. G. (1983). Unstable B chromosomes in *Silene maritima* With. (Caryophyllaceae). *Botanical Journal of the Linnean Society*, **87**, 273–283.

Crisp, P. & Jones, B. M. G. (1978). Hybridization of *Senecio squalidus* and *S. viscosus* and introgression of genes from diploid into tetraploid *Senecio* species. *Annals of Botany*, **42**, 937–944.

Deumling, B. & Greilhuber, J. (1982). Characterization of heterochromatin in different species of the *Scilla sibirica* group (Liliaceae) by in-situ hybridization of satellite DNAs and fluorochrome banding. *Chromosoma*, **84**, 535–555.

Ehrendorfer, F. (1979). Polyploidy and distribution. In: *Polyploidy, Biological Relevance* (Ed. by W. H. Lewis), pp. 45–60. Plenum Press, New York.

Elkington, T. T. (1964). *Myosotis alpestris* F. W. Schmidt, Biological Flora of the British Isles. *Journal of Ecology*, **52**, 709–722.

Elkington, T. T. (1969). Cytotaxonomic variation in *Potentilla fruticosa* L. *The New Phytologist*, **68**, 151–160.

Ellis, W. M., Calder, D. M. & Lee, B. T. O. (1970). A diploid population of *Poa annua* L. from Australia. *Experientia*, **26**, 1156.

Fearn, G. M. (1972). The distribution of intraspecific chromosome races of *Hippocrepis comosa* L. and their phytogeographical significance. *The New Phytologist*, **71**, 1221–1225.

Fearn, G. M. (1973). *Hippocrepis comosa* L.: Biological Flora of the British Isles. *Journal of Ecology*, **61**, 915–926.

Fearn, G. M. (1977). A morphological and cytological investigation of *Cochlearia* populations on the Gower Peninsula, Glamorgan. *The New Phytologist*, **79**, 455–458.

Fukuda, I. & Channell, R. B. (1975). Distribution and evolutionary significance of chromosome variation in *Trillium ovatum*. *Evolution*, **29**, 257–266.

Gadella, T. W. J. & Kliphuis, E. (1971). Cytotaxonomic studies in the genus *Symphytum*, 3. Some *Symphytum* hybrids in Belgium and the Netherlands. *Biologisch Jaarboek*, **39**, 97–101.

Gadella, T. W. J., Kliphuis, E. & Perring, F. H. (1974). Cytotaxonomic studies in the genus *Symphytum*, 6. Some notes on *Symphytum* in Britain. *Acta botanica neerlandica*, **23**, 433–437.

Gairdner, A. E. (1939). In: Maude, P. F. The Merton Catalogue. A list of the chromosome numbers of species of British flowering plants. *The New Phytologist*, **38**, 1–31.

Gill, J. A. & Davy, A. J. (1983). Variation and polyploidy within lowland populations of the *Betula pendula/B. pubescens* complex. *The New Phytologist*, **94**, 433–451.

Gill, B. S. & Kimber, G. (1974). Giesma C-banding and the evolution of wheat. *Proceedings National Academy of Science of United States of America*, **71**, 4086–4090.

Gill, J. J. B. (1965). Diploids in the genus *Cochlearia*. *Watsonia*, **6**, 188–189.

Gill, J. J. B. (1971a). Cytogenetic studies in *Cochlearia*. The chromosomal homogeneity within both the $2n = 12$ diploids and the $2n = 14$ diploids and the cytogenetic relationships between the two chromosome levels. *Annals of Botany* **35**, 947–956.

Gill, J. J. B. (1971b). The cytology and transmission of accessory chromosomes in *Cochlearia pyrenaica* DC. (Cruciferae). *Caryologia*, **24**, 173–181.

Gill, J. J. B. (1973). Cytogenetic studies in *Cochlearia* L. (Cruciferae). The origins of *C. officinalis* L. and *C. micacea* Marshall. *Genetica*, **44**, 217–234.

Gill, J. J. B. (1976). Cytogenetic studies in *Cochlearia* L. (Cruciferae). The chromosomal constitution of *C. danica* L. *Genetica*, **46**, 115–127.

Gill, J. J. B., Jones, B. M. G., Marchant, C. J., McLeish, J. & Ockendon, D. J. (1972). The distribution of chromosome races of *Ranunculus ficaria* L. in the British Isles. *Annals of Botany*, **36**, 31–47.

Gill, J. J. B., McAllister, H. A. & Fearn, G. M. (1978). Cytotaxonomic studies on the *Cochlearia officinalis* L. group from inland stations in Britain. *Watsonia*, **12**, 15–21.

Grau, J. (1964). Die Zytotaxonomie der *Myosotis alpestris* – und der *Myosotis silvatica* Gruppe in Europa. *Österreichische botanische Zeitschrift*, **111**, 561–617.

Greilhuber, J., Deumling, B. & Speta, F. (1981). Evolutionary aspects of chromosome banding, heterochromatin, satellite DNA, and genome size in *Scilla* (Liliaceae). *Berichte der Deutschen botanischen Gesellschaft*, **94**, 249–266.

Grime, J. P. (1983). Prediction of weed and crop response to climate based upon measurements of nuclear DNA content. *Aspects of Applied Biology* 4. Influence of environmental factors on herbicide performance and crop and weed biology, pp. 87–98.

Hagerup, O. (1927). *Empetrum hermaphroditum* (Lge.) Hagerup, a new tetraploid, bisexual species. *Dansk Botanisk Arkiv*, **5**, 1–17.

Hancock, B. L. (1942). Cytological and ecological notes on some species of *Galium* L. em. Scop. *The New Phytologist*, **41**, 70–78.

Harlan, J. R. & De Wet, J. M. J. (1975). On Ö. Winge and a prayer: the origins of polyploidy. *Botanical Review*, **41**, 361–390.

Harley, R. M. (1975). *Mentha*. In: *Hybridization and the Flora of the British Isles* (Ed. by C. A. Stace), pp. 383–390. Academic Press, London.

Harley, R. M. & Brighton, C. A. (1977). Chromosome numbers in the genus *Mentha*. *Botanical Journal of the Linnean Society*, **74**, 71–96.

Heslop-Harrison, J. (1953). The North American and Lusitanian elements in the Flora of the British

Isles. In: *The Changing Flora of Britain* (Ed. by J. E. Lousley), pp. 105–123. Botanical Society of the British Isles, Oxford.

HESLOP-HARRISON Y. (1953). Cytological studies in the genus *Rubus*, I. Chromosome numbers in the British *Rubus* flora. *The New Phytologist*, **52**, 22–39.

HOLMES, D. W. & BOUGOURD, S. M. (1983). B-chromosomes in *Allium schoenoprasum*. In: *Kew Chromosome Conference* vol. II (Ed. by P. E. Brandham & M. D. Bennett), p. 347. G. Allen & Unwin, London.

HUMPHREYS, M. O. (1975). The evolutionary relationships of British species of *Papaver* in section Orthorhoeades as shown by observations on interspecific hybrids. *The New Phytologist*, **74**, 485–493.

INGRAM, R. (1967). On the identity of the Irish populations of *Sisyrinchium*. *Watsonia*, **6**, 283–289.

JACKSON, R. C. (1962). Interspecific hybridisation in *Haplopappus* and its bearing on chromosome evolution in the Blepharodon section. *American Journal of Botany*, **49**, 119–132.

JONES, J. D. G. & FLAVELL, R. B. (1982). The mapping of highly-repeated DNA families and their relationship to C-bands in chromosomes of *Secale cereale*. *Chromosoma*, **86**, 595–612.

JONES, K. (1958). Cytotaxonomic studies in *Holcus* I. The chromosome complex in *Holcus mollis* L. *The New Phytologist*, **57**, 191–210.

JONES, K. & CARROLL, C. P. (1962). Cytotaxonomic studies in *Holcus*. II. Morphological relationships in *Holcus mollis* L. *The New Phytologist*, **61**, 63–71.

JONES, R. N. (1975). B-chromosome systems in flowering plants and animal species. *International Review of Cytology*, **40**, 1–100.

JORDAN, A. (1852). *Pugillus Plantarum novarum praesertim Gallicarum*, pp. 9–12. Paris.

JORDAN, A. (1864). *Diagnoses d'espèces nouvelles ou méconnues pour servir de Matériaux a une Flore refermee de la France et des Contrées voisines*, pp. 207–249. Paris.

KAY, Q. O. N. (1969). The origin and distribution of diploid and tetraploid *Tripleurospermum inodorum* (L.) Schultz Bip. *Watsonia*, **7**, 130–141.

KLIPHUIS, E. (1974). Cytotaxonomic studies in *Galium palustre* L. *Proceedings. Koniklije Nederlandse akadamie van Wetenschappen (Biol. Med.)*, **77**, 408–425.

KOOTIN-SANWU, M. & WOODELL, S. R. J. (1969). Supernumary chromosomes in *Caltha palustris*. In: *Chromosomes Today*, vol. 2 (Ed. by C. D. Darlington & K. R. Lewis). Supplement to *Heredity* **24**, pp. 192–196.

LORD, R. M. & RICHARDS A. J. (1977). A hybrid swarm between the diploid *Dactylorhiza fuchsii* (Druce) Soo and the tetraploid *D. purpurella* (T. & T. A. Steph.) Soo in Durham. *Watsonia*, **11**, 205–210.

MARCHANT, C. J. (1967). Evolution in *Spartina* (Gramineae), 1. The history and morphology of the genus in Britain. *Journal of the Linnean Society, Botany*, **60**, 1–24.

MARCHANT, C. J. (1968). Evolution in *Spartina* (Gramineae), 2. Chromosomes, basic relationships and the problem of *S.* × *townsendii* agg. *Journal of the Linnean Society, Botany*, **60**, 381–409.

MARCHANT, C. J. & BRIGHTON, C. A. (1974). Cytological diversity and triploid frequency in a complex population of *Ranunculus ficaria* L. *Annals of Botany*, **38**, 7–15.

MOORE, D. M. (1978). The chromosomes and plant taxonomy. In: *Essays in Plant Taxonomy* (Ed. by H. E. Street), pp. 39–56. Academic Press, London.

MOORE, D. M. & HARVEY M. J. (1961). Cytogenetic relationships of *Viola lactea* Sm. and other west European arosulate violets. *The New Phytologist*, **60**, 85–95.

MURRAY, B. G. (1976a). The cytology of the genus *Briza* L. (Gramineae), 3. B chromosomes. *Chromosoma*, **59**, 73–81.

MURRAY, B. G. (1976b). The cytology of the genus *Briza* L. (Gramineae), 2. Chiasma frequency, polyploidy and interchange heterozygosity. *Chromosoma*, **57**, 81–93.

NICHOLSON, G. G. (1983). Studies on the distribution and the relationship between the chromosome races of *Ranunculus ficaria* L. in S.E. Yorkshire. *Watsonia*, **14**, 321–328.

OCKENDEN, D. J. (1968). Biosystematic studies in the *Linum perenne* group. *The New Phytologist*, **67**, 787–813.

OCKENDEN, D. J. (1971). Taxonomy of the *Linum perenne* group in Europe. *Watsonia*, **8**, 205–235.

PACKHAM, J. R. (1983). *Lamiastrum galeobdolon* (L.) Ehrend. & Polatschek: Biological Flora of the British Isles. *Journal of Ecology*, **71**, 975–997.

PARKER, J. S. (1975). Aneuploidy and isolation in two *Hypochaeris* species. *Chromosoma*, **52**, 89–101.

PERRING, F. (1969). Network Research: *Symphytum* survey. *Proceedings Botanical Society of the British Isles*, **7**, 553–556.

PRICE, H. J., CHAMBERS, K. L. & BACHMANN, K. (1981). Geographic and ecological distribution of genomic DNA content variation in *Microseris douglasii* (Asteraceae). *Botanical Gazette*, **142**, 415–426.

RICHARDS, A. J. (1973). The origin of *Taraxacum* agamospecies. *Botanical Journal of the Linnean Society*, **66**, 189–211.

ROSSER, E. M. (1955). A new British species of *Senecio*. *Watsonia*, **3**, 228–232.

SCHWEIZER, D. (1979). Fluorescent chromosome banding in plants: applications, mechanisms and implications for chromosome structure. In: *The Plant Genome;* Proceedings of the 4th John Innes Symposium, Norwich (Ed. by D. R. Davies & R. A. Hopwood), pp. 61–72.

SEAVEY, S. R. & RAVEN, P. H. (1977a). Chromosomal evolution in *Epilobium* sect. *Epilobium* (Onagraceae). *Plant Systematics & Evolution*, **127**, 107–119.

SEAVEY, S. R. & RAVEN, P. H. (1977b). Chromosomal evolution in *Epilobium* sect. *Epilobium* (Onagraceae), 2. *Plant Systematics & Evolution*, **128**, 195–200.

SHERMAN, M. (1945). Karyotype evolution: a cytogenetic study of seven species and six interspecific hybrids of *Crepis*. *University of California Publications in Botany*, **18**, 369–408.

SIEBER, V. K. & MURRAY, B. G. (1980). Spontaneous polyploids in marginal populations of *Alopecurus bulbosus* Gouan (Poaceae), *Botanical Journal of the Linnean Society*, **81**, 293–300.

SIEBER, V. K. & MURRAY, B. G. (1981). Structural and numerical chromosomical polymorphism in natural populations of *Alopecurus* (Poaceae). *Plant Systematics & Evolution*, **139**, 121–136.

SMITH, G. L. (1963a). Studies in *Potentilla* L. I. Embryological investigations into the mechanism of agamospermy in British *P. tabernaemontani* Aschers. *The New Phytologist*, **62**, 264–282.

SMITH, G. L. (1963b). Studies in *Potentilla* L. II. Cytological aspects of apomixis in *P. crantzii* (Cr.) Beck ex Fritsch. *The New Phytologist*, **62**, 283–360.

SMITH, G. L., BOZMAN, V. G. & WALTERS, S. M. (1971). Studies in Potentilla L. III. Variation in British *P. tabernaemontani* Aschers. and *P. crantzii* (Cr.) Beck ex Fritsch. *The New Phytologist*, **70**, 607–618.

STACE, C. A. (1975). Introductory. In: *Hybridization and the Flora of the British Isles* (Ed. by C. A. Stace), pp. 1–90. Academic Press, London.

STACE, C. A. (1977). The origin of radiate *Senecio vulgaris* L. *Heredity*, **39**, 383–388.

STACE, C. A. (1980). *Plant Taxonomy and Biosystematics*. Edward Arnold, London.

STEBBINS, G. C. (1971). *Chromosomal Evolution in Higher Plants*. Edward Arnold, London.

TEPPNER, H., EHRENDORFER, F. & PUFF, C. (1976). Karyosystematic notes on the *Galium palustre*-group (Rubiaceae). *Taxon*, **25**, 95–97.

TIMMS, E. W. & CLAPHAM, A. R. (1940). Jointed rushes of the Oxford District. *The New Phytologist*, **39**, 1–16.

TUTIN, T. G. (1957). A contribution to the experimental taxonomy of *Poa annua* L. *Watsonia*, **4**, 1–10.

VOSA, C. G. (1973a). Heterochromatin recognition and analysis of chromosome variation in *Scilla sibirica*. *Chromosoma*, **43**, 269–278.

VOSA, C. G. (1973b). Quinacrine fluorescence analysis of chromosome variation in the plant *Tulbaghia leucantha*. *Chromosomes Today*, vol. 4 (Ed. by J. Wahrman & K. R. Lewis), pp. 345–349. J. Wiley, New York; Israel Universities Press, Jerusalem.

VOSA, C. G. (1975). The use of Giemsa and other staining techniques in karyotype analysis. *Current Advances in Plant Science*, **6**, 495–510.

VOSA, C. G. (1976a). Heterochromatic patterns in *Allium*. I. The relationship between the species of the *Cepa* group and its allies. *Heredity*, **36**, 383–392.

VOSA, C. G. (1976b). Heterochromatic banding patterns in *Allium* II. Heterochromatin variation in the species of the Paniculatum group. *Chromosoma*, **57**, 119–133.

WARBURG, E. F. (1962). *Sorbus*. In: *Flora of the British Isles* by Clapham, A. R., Tutin, T. G. & Warburg, E. F. 2nd ed. pp. 423–437. University Press, Cambridge.

WEGMÜLLER, S. (1971). A cytotaxonomic study of *Lamiastrum galeobdolon* (L.) Ehrend. & Polatschek in Britain. *Watsonia*, **8**, 277–288.

WEIR, J. & INGRAM, R. (1980). Ray morphology and cytological investigations of *Senecio cambrensis* Rosser. *The New Phytologist*, **86**, 237–241.

WINGE, Ö. (1940). Taxonomic and evolutionary studies in *Erophila* based on cytogenetic investigations. *Comptes rendus des séances de la Laboratoire Carlsberg (Serie Physiologie)*, **23**, 41–74.

YEO, P. F. (1956). Hybridization between diploid and tetraploid species of *Euphrasia*. *Watsonia*, **3**, 253–269.

ZANDEE, M. (1981). Studies in the *Juncus articulatus* L. – *J. acutiflorus* Ehrh. – *J. anceps* Laharpe – *J. alpinus* Vill. aggregate. *Proceedings. Koniklijke Nederlandse akadamie von Wetenschappen C*, **84**, 243–254.

ZANDEE, M. & SCHEEPEN, J. VAN. (1981). Studies in the *Holcus lanatus-Holcus mollis* complex (Poaceae: Gramineae), 1. A cytotaxonomic survey. *Proceedings. Koniklijke Nederlandse akadamie von Wetenschappen C*, **84**, 479–491.

THE FLORA AND VEGETATION OF BRITAIN: ECOLOGY AND CONSERVATION

By C. D. PIGOTT

Department of Biological Sciences, University of Lancaster, LA1 4YQ UK

Summary

A main aim of studies of the British flora in future must be ecological, with plant ecology developed as a multidisciplinary and experimental science which has as its primary purpose the understanding of vegetation. The response of species to environment is largely mediated through the response of vegetation. It is essential to have an accepted and widely understood classification of British vegetation.

Destruction of the more natural types of vegetation, with accompanying loss of species, should be a matter of national concern as many communities are irreplaceable. Their loss represents irreversible damage to our cultural heritage and a deterioration of the quality of our environment. Ecological understanding of vegetation is necessary to ensure that conservation achieves its primary purpose.

Key words: British vegetation, British flora, *Acer pseudoplatanus*, conservation, ecology.

Introduction

The contributions to this symposium illustrate several very different ways of studying the British flora but each reflects a deeply held interest of one person. Professor A. R. Clapham is one of the authors of the definitive flora of the British Isles (Clapham, Tutin & Warburg, 1962) and of the excursion flora (Clapham, Tutin & Warburg, 1981), but Dr S. M. Walters reminds us that he was also a member of the organizing committee and a contributor to *Flora Europaea* (Tutin et al., 1964–1980) which was published in Britain but is now the definitive flora for the whole continent. This might well lead future generations to regard Clapham as primarily a taxonomist, but I believe that they would be guilty of a serious error of identification, for his real interests are in the biology, physiology and ecology of plants. A sound knowledge of the flora of a region and the ability to identify species correctly is surely an essential prerequisite for the study of vegetation, whether descriptive, analytical or experimental. It becomes increasingly rare for professional botanists to have this knowledge which is more often found among amateurs.

Dr T. T. Elkington's contribution reminds us that even a knowledge of species is not enough to appreciate their full ecological potential. Many species contain more than one genetically, and sometimes cytologically, distinct race which may differ in geographical distribution and ecological characteristics. In their pioneer work on chromosome numbers and chromosome morphology, botanists made fundamental contributions to the understanding of evolution and in the early part of his career Clapham investigated two species within which there exist races with different chromosome numbers (Timm & Clapham, 1940; Hancock, 1942).

Two papers by Professor A. G. Smith and by Professor G. W. Dimbleby have discussed aspects of the history of the British flora and the influence of prehistoric man on vegetation, reminding us of Clapham's contributions to investigations of buried trackways on the Somerset levels (Clapham & Godwin, 1948), of the fen at Cothill, Berkshire (Clapham & Clapham, 1939) and to peats on Cross Fell, Cumberland (Godwin & Clapham, 1951). Investigations of this type, which are fundamentally geological, depend on pollen analysis and the accurate identification of other remains of plants. They have provided a detailed knowledge of the history of vegetation since the last glaciation and explanations of many present-day features of the distribution of plants, of vegetation and of soils. Such studies are an indispensible tool of the ecologist; from them we know, for example, which trees are truly native and we can reconstruct the composition of our native forests. In so doing we may appreciate that almost all our vegetation now is to a greater or lesser extent modified by human intervention. Neither should we forget, as Dimbleby has made clear, that these botanical investigations have made a substantial contribution to archaeology.

The Study of Vegetation

It is, however, in Professor J. P. Grime's paper that our attention is sharply focussed on what is one of the most distinctive characteristics of Clapham's contribution to ecology. This is the set of ideas he presented in his presidential address to the British Ecological Society (Clapham, 1956) in which he advocated the need to study the biology of species, both in the field and in laboratory conditions, in order to understand the part they play in vegetation. In this paper he demonstrates what are surely essential requirements for an ecologist: a real interest in plants as living organisms and a curiosity which leads an ecologist, not simply to describe vegetation, but to ask why species occur in particular combinations to form vegetation, and to what extent these combinations are determined by the vegetation itself, by soil, by climate, by biotic factors and by the interactions of all these variables. Grime has emphasized, as did Clapham, what I may term the autoecological approach; elsewhere (Pigott, 1982) I have argued that autecological investigations must always be closely related to parallel studies of vegetation. This is because the response of species to a particular set of physical factors is often quite different when they are grown in isolation, or indeed in artificial mixtures, from when they are growing in real vegetation. Clapham (1956) himself remarks that there is an important difference between those who study vegetation in order to solve autecological problems and those who study autecology in order to understand vegetation. More recently, in his review of Ellenberg's *Vegetation Mitteleuropas mit den Alpen* (Clapham, 1983), he expresses his admiration for the way it is possible in central Europe to relate analytical and experimental studies of species and vegetation of a comprehensive and accepted classification of vegetation. He then expresses the hope that the current National Vegetation Classification will fill this gap in Britain.

The lack of an accepted classification of British vegetation has retarded the development of ecology and deprived nature conservation of an essential tool. Many analytical and experimental studies in ecology, whether of an autecological or what may be termed demographic nature, seem to be almost abstract because they are not securely anchored in the context of real vegetation. For example, Harper (1977) quotes information from Peterken (1966) which shows large differences in the mortality of seedlings of holly (*Ilex aquifolium*) in different types of woodland, yet most of the numerous examples discussed in this critical and comprehensive book on population ecology of plants have either no description of the vegetation, or, as in the studies of *Ranunculus*, only the briefest statement.

The need to relate such studies to vegetation may be illustrated by an experimental investigation of the conditions required for regeneration of another tree, *Tilia cordata*, in three different communities in adjacent parts of a large woodland on Lower Greensand and Weald clay in Surrey. At least 50 two year old seedlings were planted in each of three plots and their growth in height and their mortality is shown in Table 1. Growth rate is least in the *Betula pendula–Deschampsia*

Table 1. *Growth in height and annual extension (means and 95% confidence limits) of undamaged plants, and damage and mortality of plants of* Tilia cordata *planted in November 1977 in plots in three different communities in Leith Hill Place Wood, Surrey*

	Plant community		
	Betula pendula– Deschampsia flexuosa	Quercus robur– Pteridium aquilinum– Rubus fruticosus	Fraxinus excelsior– Acer campestre– Mercurialis perennis
Soil type	Humo-ferric podzol	Brown forest soil	Gleyed brown forest soil
Height of shoot in 1980 (cm)	12·8 ± 3·5	15·9 ± 2·6	19·5 ± 3·8
Extension of shoot in 1979 (cm)	7·5 ± 3·6	9·0 ± 2·2	14·9 ± 2·0
Plants damaged by Clethrionomys glareolus in 1980 (%)	0	16·3	91·7
Plants destroyed 1980–1983 (%)	11·9	36·7	86·7
Undamaged plants surviving in 1983 (%)	67	4	0

flexuosa community on podzolised sand, moderate in the *Quercus robur–Pteridium aquilinum–Rubus fruticosus* community on an acid brown forest soil and greatest in the *Fraxinus excelsior–Acer campestre–Mercurialis perennis* community on gley, but is inversely related to both damage and mortality. Where the seedlings grow most rapidly, most were destroyed in 3 years. The main cause of mortality is damage caused by bank voles (*Clethrionomys glareolus* Schreiber), and trapping in these sites and elsewhere shows that the density of these largely vegetarian rodents probably varies between plant communities and is related in particular to the cover provided by such species as *Rubus fruticosus* in the field-layer, and probably to the availability of particular species as food (Watts, 1968). In spite of numerous studies of population dynamics of small rodents in Britain and their well-recognized importance in destroying seedlings of trees, there is no published information on the relation between their density and types of vegetation. The reason is simply, I suspect, that there is no accepted classification of vegetation available to zoologists.

The National Vegetation Classification has now completed a classification of British woodlands, but this is not yet published. It is based on species present in all layers but is independent of other characteristics of the site, so that correlations can subsequently be tested. An example will show the value of this classification and the one selected is appropriate both in relation to Clapham (1956) and the paper of Walters in this symposium. It shows (Table 2) a group of woodland communities

Table 2. *Constancy (I–V: I = 1–20%, II = 21–40%, etc.) and range of Domin values of cover/abundance (1–10) of selected species in five sub-communities of the* Quercus robur–Pteridium aquilinum–Rubus fruticosus *woodland community and the overall values for the community. Preferential species for each sub-community enclosed in boxes (National Vegetation Classification).*

	a	b	c	d	e	
Tree layer						
Quercus robur	IV (3–10)	III (2–10)	IV (2–10)	IV (1–10)	II (1–8)	IV (1–10)
Betula pendula	III (2–8)	III (2–9)	I (1–8)	III (1–10)	I (1–8)	II (1–10)
Quercus petraea	—	III (3–10)	I (3)	I (3–9)	I (3–9)	II (3–10)
Fagus sylvatica	I (3)	II (1–10)	II (1–10)	I (3–5)	I (1–6)	I (1–10)
Castanea sativa	IV (3–10)	I (1–5)	I (3–7)	I (3–5)	I (2–4)	I (1–10)
Pinus sylvestris	I (4)	II (3–4)	I (3–4)	—	I (1–10)	I (1–10)
Pinus nigra var. *maritima*	—	—	—	—	I (10)	I (6–10)
Pseudotsuga menziesii	—	—	—	II (2–10)	—	I (6–10)
Acer pseudoplatanus	I (5)	II (1–9)	I (5–8)	II (6–10)	III (1–7)	II (1–9)
Fraxinus excelsior	II (2–7)	I (1–6)	II (1–7)	II (6–10)	III (1–8)	II (1–8)
Ulmus glabra	—	I (1–4)	—	I (3–9)	II (1–7)	I (1–7)
Quercus hybrids	—	I (1–8)	—	I (1–6)	II (1–10)	I (1–10)
Shrub layer						
Corylus avellana	III (2–9)	III (1–9)	IV (1–10)	I (3–7)	III (1–9)	III (1–10)
Crataegus monogyna	I (3–7)	II (1–6)	II (1–5)	I (4–6)	II (1–5)	II (1–7)
Ilex aquifolium	I (2)	II (1–6)	II (2–9)	I (3–7)	II (1–6)	II (1–9)
Field and ground layers						
Pteridium aquilinum	III (2–7)	IV (1–9)	IV (1–10)	V (1–10)	III (1–8)	IV (1–10)
Rubus fruticosus agg.	IV (2–9)	V (3–10)	V (1–10)	IV (1–10)	III (1–8)	IV (1–10)
Lonicera periclymenum	IV (3–7)	III (2–8)	V (1–8)	III (1–7)	II (1–8)	IV (1–8)
Hyacinthoides nonscripta	IV (4–10)	III (3–9)	II (1–10)	I (1–7)	III (1–9)	III (1–10)
Anemone nemorosa	III (3–6)	I (1–8)	I (3–8)	—	I (1–6)	I (1–8)
Lamiastrum galeobdolon	II (2–5)	I (1–5)	I (1–5)	—	—	I (1–5)
Hedera helix	II (2–8)	II (2)	IV (2–10)	I (2–4)	I (2–7)	I (2–10)
Holcus lanatus	I (3–9)	I (1–6)	I (1–7)	IV (1–8)	I (1)	I (1–9)
Oxalis acetosella	I (2–3)	I (1–4)	I (2)	I (2)	IV (1–9)	II (1–9)
Holcus mollis	I (2–8)	II (1–10)	II (2–10)	I (3–9)	IV (1–9)	II (1–10)
Dryopteris dilata	I (2–5)	II (1–7)	II (1–6)	I (1–5)	III (1–8)	II (1–8)

* Key: a, *Castanea sativa* coppice subcommunity; b, Typical subcommunity; c, *Hedera helix* subcommunity; d, *Holcus lanatus* subcommunity; e, *Acer pseudoplatanus–Oxalis acetosella* subcommunity.

where *Hyacinthoides nonscripta* is normally characteristic in the field-layer, and often dominant in the vernal phase. This feature is restricted to the British Isles, south Belgium and northern France (Noirfalise, 1969) so that, in the European context, these communities have high priority for conservation. Various interesting features can be seen in Table 2. Replacement of the canopy by *Castanea sativa* has little effect on the dominance of *Hyacinthoides* but such woods usually have more *Anemone nemorosa* and *Lamiastrum galeobdolon*. This is not necessarily because of the change of canopy (though critical experimental evidence is lacking) but probably because *Castanea* favours slightly moister soils with a higher proportion of clay, or a well-developed argillic B-horizon, or clay pan. Dense coppice of *Castanea*, with uniform age of stems, results, as expected, in a marked reduction of the field layer.

Replacement of the deciduous canopy by conifers results in a severe reduction of the field-layer and, when plantations reach an age of 15 to 25 years, there is almost complete suppression, as is shown by the decrease in constancy of *Hyacinthoides*. Both *Hyacinthoides* and *Anemone* spread very slowly into areas of woodland from which they have been excluded. Rates of 6 to 10 m and 1 to 2 m in a century, respectively, have been measured in this type of woodland in Surrey, so that if eliminated by conifers, a subsequent rotation of oak will not give sufficient time for the re-establishment of their dominance.

Another feature of importance, which can only be recognized from the vegetation, concerns sycamore (*Acer pseudoplatanus*). Although introduced into Britain in the Middle Ages (Jones, 1944), this tree now occurs in woodlands almost throughout the British Isles (Fig. 1) where it regenerates from its prolific fertile fruit. It is a species of considerable diagnostic value in the classification of woodlands and surely should not be neglected because it is alien. It is present in all five subcommunities in Table 2, but is most frequent and often abundant in the *Acer pseudoplatanus–Oxalis acetosella* subcommunity. This is predominantly a type of woodland of the north and west of Britain and contrasts with the other subcommunities (Fig. 2). The behaviour of sycamore is entirely in keeping with its natural distribution in the hills of the Ardennes, Vosges, Black Forest and Alps, and its rarity in the drier parts of northern France. A possible factor controlling the frequency of sycamore is seen when the distribution of communities in which the species has a high frequency is plotted on the same map as the 30 inch (762 mm) isohyet for mean annual rainfall (Fig. 3). This relationship emerges only from a study of sycamore in vegetation and not from its overall distribution (Fig. 1).

Correlations are not proof of causality and experimental work on the water-relations of sycamore is now in progress at Lancaster by Dr W. J. Davies and Miss G. Taylor. If supply of water proves to be critical, then it is relevant to draw attention to the work of Whittaker (1984) on mature trees which shows that resistance to loss of water from leaves is significantly reduced by damage caused by the widespread and locally common Typhlocybine leaf-hopper (*Ossiannilssonola callosa* Then.).

I hope these examples provide sufficient evidence to satisfy those experimental ecologists who doubt the value of classification of vegetation or are even hostile to its purpose. I would argue that the study of vegetation is the essence of plant ecology and that it is difficult to study vegetation without a system of reference to communities.

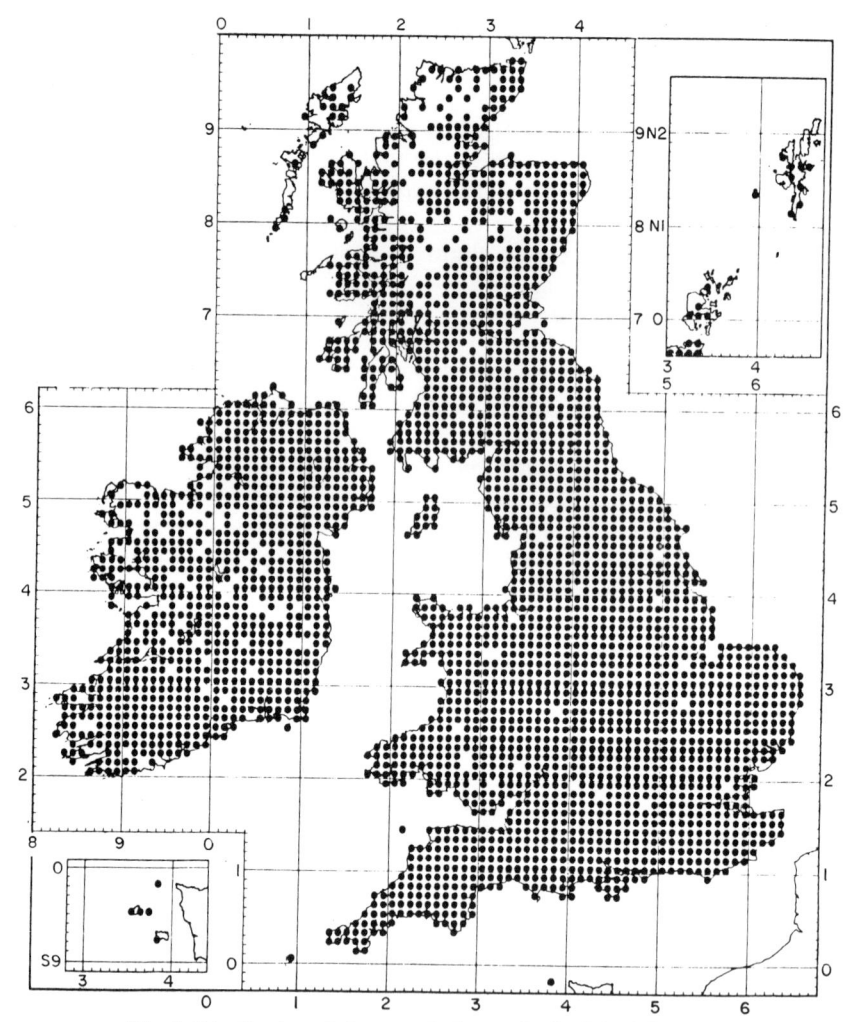

Fig. 1. Distribution of *Acer pseudoplatanus* in the British Isles.
(from Perring & Walters, 1962).

Plant Ecology as a Multidisciplinary Science

My examples have also been selected to emphasize the multidisciplinary nature of plant ecology. There is a need to be able to identify species accurately, to understand their genetic variability and to recognize types of vegetation. Then, in seeking explanations for the patterns which emerge, ecology demands an understanding of some parts of meteorology, soil science, plant physiology and biochemistry, and sufficient zoology to recognize the interrelationships between plants and both vertebrates and invertebrates. I have also referred to the considerable inertia in some types of vegetation: conditions may change so rapidly that many species have had insufficient time to occupy all the situations where they would succeed were they to be introduced, so that a knowledge of the history and even the prehistory of vegetation is also important.

Ecologists cannot predict to which discipline they will need to turn in order to

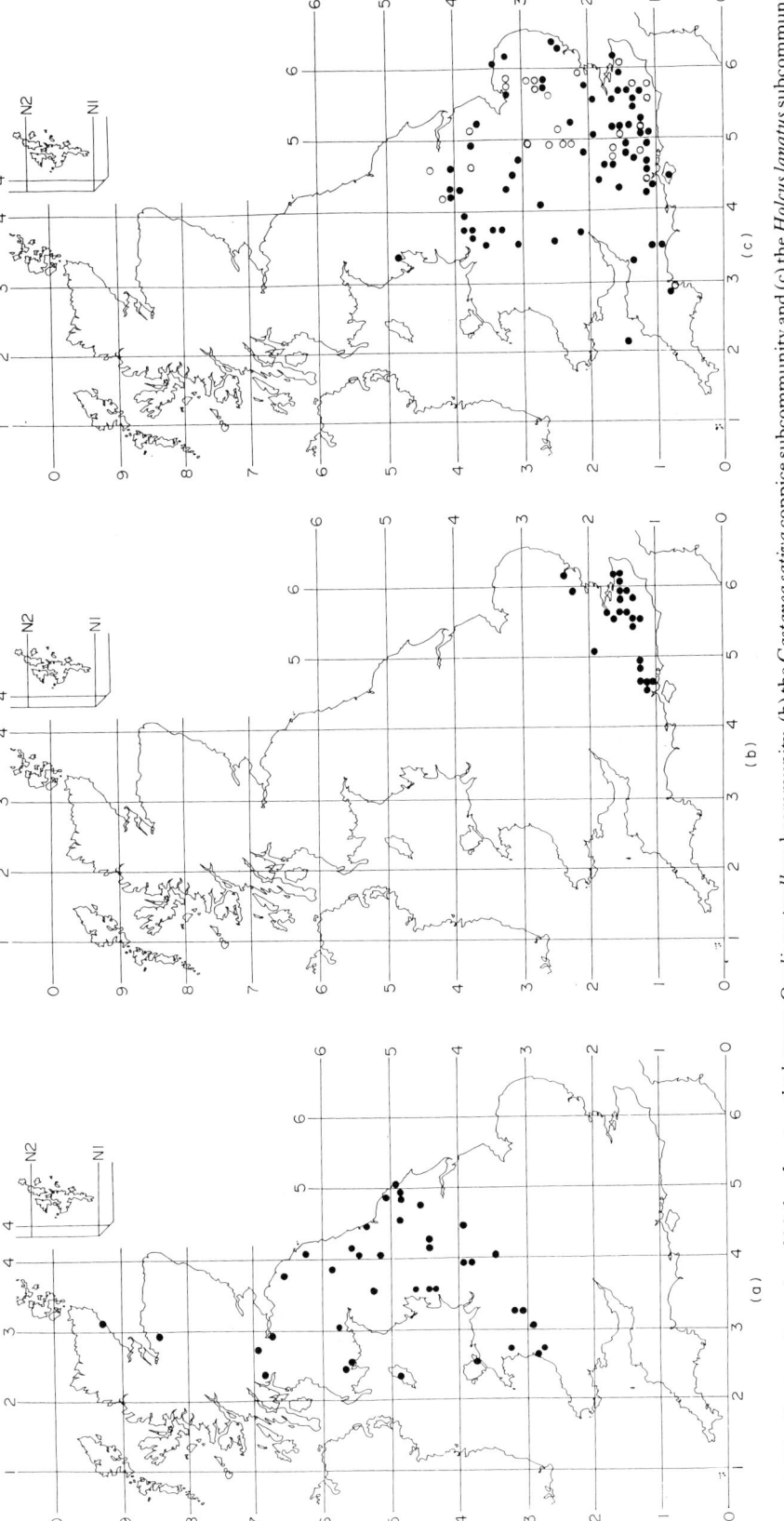

Fig. 2. Distribution of samples of (a) the *Acer pseudoplatanus–Oxalis acetosella* subcommunity, (b) the *Castanea sativa* coppice subcommunity and (c) the *Holcus lanatus* subcommunity (including plantations of conifers indicated by open circles) of the *Quercus robur–Pteridium aquilinum–Rubus fruticosus* community.

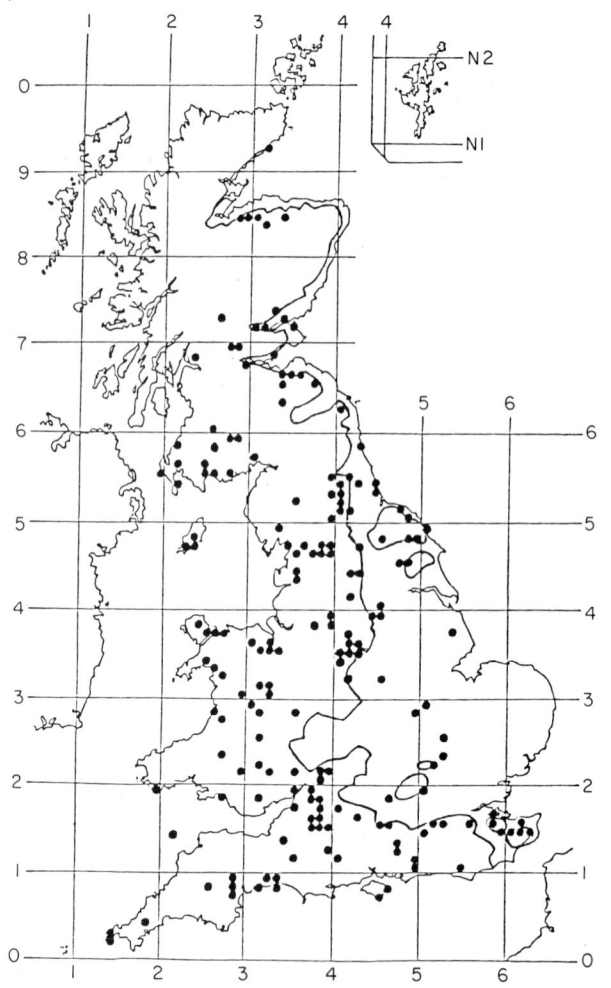

Fig. 3. Distribution of samples of woodland communities in which *Acer pseudoplatanus* has a constancy of III or more in relation to the 762 mm (30 inch) isohyet of mean annual rainfall.

solve their problems; they must retain wide interests and resist the temptation to become narrowly specialized.

Ecology Applied to Conservation

An appreciation of the inertia of vegetation is fundamental to recognizing the need for conservation of vegetation, not only in the densely populated, industrialized countries such as Britain, but also worldwide. It is therefore not surprising that a person with Clapham's wide interests in our native flora should be sincerely committed to the work of the Nature Conservancy, now the Nature Conservancy Council, and should have served on its Scientific Policy Committee, and been both a member and chairman of its council.

Much of the vegetation of Britain and its associated communities of animals have been shown to be the product of complex interrelationships and of long-continued change. Evidence of how these ecosystems function is hidden within them, in the vegetation, the animal populations and the soils. At present we know only a small part of what there is to learn. If the systems are destroyed, the information they contain is lost for ever. As Ashby (1979) has argued in his search for an ethical basis for conservation, although some destruction is inevitable, every effort should be made to avoid destruction of anything which is irreplaceable. It is often suggested that vegetation can always be recreated, and this may well be true for some types of vegetation which have developed recently from certain human activities. But for many communities the replica would never be the same as the original. To take an extreme example, an area of woodland, dominated by oak, elm and lime, which can be demonstrated to have had continuity with that forest which covered much of England and Wales before BC 3000 can never be recreated because we simply do not know about all the features associated with five millenia of existence. It is only recently that Rackham (1980) has shown that such woodlands are store-houses of historical information. Deposits of loess are retained in the soils but have been lost from surrounding agricultural surfaces (Ball & Stevens, 1981). Even the chromosome number of one of the most characteristic trees, *Tilia cordata*, has not been determined for plants from native populations. A horticultural attitude to conservation would not achieve the primary purpose of nature conservation. A parallel may be drawn from the conservation of archaeological sites and ancient buildings. Many of the buildings which were destroyed in the war in such cities as Warsaw and Dresden have since been replaced by replicas. Such replicas can provide no more information than was available when they were recreated. Medieval frescos hidden by layers of plaster cannot be discovered, nor can the proportion of radioactive carbon in the timbers have its original meaning.

In Dr D. A. Ratcliffe's paper, we are confronted with data showing the acceleration of change which now affects our native vegetation and which often leads to the total destruction of woodlands, heaths, pastures, meadows, bogs and fens. This change is a proceeding on a vast scale. Should we be concerned at the loss? I am not one who believes that scientific discoveries of great practical value to agriculture will come from the study of native British plants, although it should not be forgotten that several of our native trees provide valuable timber; and forestry in Britain has, I believe, much to gain from a more ecological attitude. The true value of the British flora is surely its place in our culture and our environment. The destruction of our natural vegetation is nothing less than the destruction of our cultural heritage, so that, for example, alien conifers and southern beech (*Nothofagus* spp.) are now extensively planted in preference to our native species of oak. The consequence is a loss of the characteristic associated species of plants and animals. Woods, which in May are blue with *Hyancinthoides nonscripta* and heavy with its scent, should be available for everyone to enjoy and it is absurd to suppose that the real benefits they offer to us will be adequately provided by a few widely scattered nature reserves.

The justification in practical terms for the scientific study of the British flora is that this alone provides an understanding of what must be done to conserve species within ecosystems which are as natural as possible. The quality of our environment is at stake and this surely deserves the expenditure of some public money to support research, so that the 142 215 ha of land protected by the Nature Conservancy Council (1984), or the 201 466 ha (including over 700 km of coast) protected by the National Trust can be maintained in ways which ensure the survival of an attractive landscape containing as much as possible of our native flora and fauna.

Acknowledgements

I thank Dr John Rodwell for preparing the information from the National Vegetation Classification and for his permission to refer to his work on the distribution of sycamore.

References

Ashby, E. (1979). The search for an environmental ethic. *The Tanner Lecture on Human Values*, University of Utah.
Ball, D. F. & Stevens, P. A. (1981). The role of 'ancient' woodlands in conserving 'undisturbed' soils in Britain. *Biological Conservation*, **19**, 163–176.
Clapham, A. R. (1965). Autecological studies and the 'Biological Flora of the British Isles'. *Journal of Ecology*, **44**, 1–12.
Clapham, A. R. (1983). Review of Ellenberg, H., 1982. *The New Phytologist*, **93**, 153–154.
Clapham, A. R. & Clapham, B. N. (1939). The valley fen at Cothill, Berkshire. *The New Phytologist*, **38**, 167–174.
Clapham, A. R. & Godwin, H. (1948). Studies on the post-glacial history of British vegetation 8. Swamping surfaces in peats of the Somerset Levels; 9. Prehistoric trackways in the Somerset Levels. *Philosophical Transactions of the Royal Society* B, **233**, 233–249, 249–273.
Clapham, A. R., Tutin, T. G. & Warburg, E. F. (1962). *Flora of the British Isles*. Cambridge University Press, Cambridge.
Clapham, A. R., Tutin, T. G. & Warburg, E. F. (1981). *Excursion Flora of the British Isles*. Cambridge University Press, Cambridge.
Ellenberg, H. (1982). *Vegetation Mitteleuropas mit den Alpen*. Ulmer, Stuttgart.
Godwin, H. & Clapham, A. R. (1951). Peat deposits on Cross Fell, Cumberland. *The New Phytologist*, **50**, 167–171.
Hancock, B. L. (1942). Cytological and ecological notes on some species of *Galium* L. em Scop. *The New Phytologist*, **41**, 70–78.
Harper, J. L. (1977). *Population Biology of Plants*. Academic Press, London.
Jones, E. W. (1944). Biological Flora of the British Isles: *Acer* L. *Journal of Ecology*, **32**, 215–252.
Nature Conservancy Council (1984). *Ninth Annual Report of the Nature Conservancy Council*.
Noirfalise, A. (1969). La chênaie melangée à jacinthe du domaine Atlantique de l'Europe (Endymio-Carpinetum). *Vegetatio: Acta geobotanica*, **17**, 131–150.
Perring, F. H. & Walters, S. M. (1962). *Atlas of the British Flora*. Thomas Nelson, London.
Peterken, G. F. (1966). Mortality of holly (*Ilex aquifolium*) seedlings in relation to natural regeneration in the New Forest. *Journal of Ecology*, **54**, 259–269.
Pigott, C. D. (1982). The experimental study of vegetation. *The New Phytologist*, **90**, 389–404.
Rackham, O. (1980). *Ancient Woodland: Its History, Vegetation and Uses in England*. Edward Arnold, London.
Timm, E. W. & Clapham, A. R. (1940). Jointed rushes of the Oxford district. *The New Phytologist*, **39**, 1–16.
Tutin, T. G., Heywood, V. H., Burges, N. A., Valentine, D. H., Webb, D. A. (1964–1980). *Flora Europaea*. Cambridge University Press, Cambridge.
Watts, C. H. S. (1968). The foods eaten by wood mice (*Apodemus sylvaticus*) and bank voles (*Clethrionomys glareolus*) in Wytham Woods, Berkshire. *Journal of Animal Ecology*, **37**, 25–41.
Whittaker, J. B. (1984). Responses of sycamore (*Acer pseudoplatanus*) leaves to damage by a Typhlocybine leaf hopper, *Ossiannilssonola callosa*. *Journal of Ecology*, **72**, 455–462.

INDEX

Agriculture 57–72
Agriculture: development 78–84
Alder 35–55
Angiosperm families 15–33
Anthropogenic changes from neolithic through medieval times 57–72
Atlas Florae Europaeae 9–12
Boreal–Atlantic transition, and Newferry 35–55
British flora, contemporary: the ecology of species, families and communities 15–33
British flora: cytogenetic variation, origins and significance 101–118
British flora: ecology and conservation 119–128
British flora: relation to European floras 3–13
British vascular plants 3–13
British vegetation: ecology and conservation 119–128
British vegetation, post-medieval and recent changes: culmination of the human influence 73–100
Bronze Age: vegetative change 67–69
B chromosomes 111–113
Clapham, A. R. 1, 3, 13, 57, 119–120
 conservation 126
 cytogenic research 101, 103, 119
 Distribution Maps Scheme 4, 9, 77
 ecological study of species 15, 17, 25, 119–120
 European view 4
 Flora Europaea 1, 4, 9, 119
Climatic fluctuation 74–76
Communities of contemporary British flora: ecology 25–28
Competitive plants 29–33
Conservation: flora and vegetation of Britain 95–98, 119–128
Cytogenetic variation in the British flora: origins and significance 101–118
Dimbleby, G. W. Anthropogenic changes from neolithic through medieval times 57
Distribution Maps Scheme 4, 9
Distribution studies 3–13
Diversity 15–33
Dot distribution maps 3–13
Ecology and conservation: flora and vegetation of Britain 119–128
Ecology of species, families and communities of the contemporary British flora 15–33
Elkington, T. T. Cytogenetic variation in the British flora: origins and significance 101
Environment 57–72

Estuarine habitats at the Boreal–Atlantic transition 46
European floras: relation to British flora 3–13
European vascular plants 3–13
Evolution 101–118
Families of contemporary British floras: ecology 21–25
Flora, *see* British flora
Flora Europaea 1, 4, 7, 9–12, 119
Forestry: development 84–87
Godwin, Sir H. 5, 36, 37–39
Grime, J. P. The ecology of species, families and communities of the contemporary British flora 15
Harley, J. L. Introduction to Symposium 1
Human influence at the Boreal–Atlantic transition 46–50
Human influence: post-medieval and recent changes in British vegetation 73–100
Iron Age: vegetative change 69–70
Landolt, E. 8
Mapping 4, 9–12, 77
Maps, dot distribution 3–13
Matthews, J. R. 5–6, 9
Neolithic: anthropogenic changes 57–72
Neolithic: vegetative change 61–67
Newferry and the Boreal–Atlantic transition 35–55
Pigott, C. D. The flora and vegetation of Britain: ecology and conservation 119
Pollen analysis 35–55, 58, 59–60
Polyploidy 101–118
Prehistoric vegetation 57–72
Radiocarbon dating 35–55
Ratcliffe, D. A. Post-medieval and recent changes in British vegetation: the culmination of human influence 73
Recreation 92–93
Smith, A. G. Newferry and the Boreal–Atlantic transition 35
Species of contemporary British flora: ecology 15–21
Strategy 15–33
Stress-tolerant plants 29–33
Ruderal plants 29–33
Urban industrial spread 87–92
Vegetation, prehistoric 57–72
Vegetation, *see also* British vegetation
Vegetational changes, post-medieval 73–100
Vegetational changes, recent 73–100
Walters, S. M. The relation between the British and the European flora 3